Nowhere Left to Go

Nowhere Left to Go

How Climate Change Is Driving
Species to the Ends of the Earth

BENJAMIN VON BRACKEL

Translated by Ayça Türkoğlu

THE EXPERIMENT

NEW YORK

The Experiment, LLC
220 East 23rd Street, Suite 600
New York, NY 10010-4658
theexperimentpublishing.com

The Experiment's books are available at special discounts when purchased in bulk for premiums and sales promotions as well as for fund-raising or educational use. For details, contact us at info@theexperimentpublishing.com.

The translation of this work was supported by a grant from the Goethe-Institut.

Library of Congress Cataloging-in-Publication Data available upon request

ISBN 978-1-61519-861-0
Ebook ISBN 978-1-61519-862-7

Jacket and text design, and original maps, by Jack Dunnington
Author photograph © Stöhr Fotografie

Manufactured in the United States of America

First printing July 2022
10 9 8 7 6 5 4 3 2 1

For Oliv

Contents

III

EXODUS IN THE TROPICS

IV

SOLUTIONS

Edith's Checkerspot, of the subspecies quino *(top) and* beani *(bottom), is a captive of its environment, unable to survive if the climate grows too warm or arid.*

It Begins

Southern California, before the turn of the millennium

It couldn't have been more innocent to begin with. In the San Ysidro Mountains close to the Mexican border, a butterfly—a checkerspot—spreads its wings to reveal a flecked pattern in red and black. It takes off, before being caught in a gust of wind and borne hundreds of feet up the mountain.

Its fate seems sealed the moment it lands. For thousands of years, countless others of its species followed the same route, a chance gust blowing them up toward the summit. All have died without leaving any offspring. Evolution has specifically ensured that *Euphydryas editha*, or the Edith's checkerspot, can thrive only within a narrow temperature window. Should this species stray too far outside its usual climate, it will not survive.

Yet for this checkerspot—a female—something amazing is going to happen: She is going to live. Using the olfactory organs on her legs and feelers, she will sniff out wildflowers in this unfamiliar landscape and lay dozens of eggs, which she has carried around for weeks, on the undersides of their leaves. Then, the next stages of metamorphosis will begin; the caterpillars will hatch, eat, and pupate before emerging as butterflies.

She will establish a new colony.

But why has *Euphydryas editha* failed to colonize this higher spot before when this little checkerspot has succeeded? She is no more cunning, no stronger or better able to adapt, than those that came before her.

If this checkerspot has not undergone any fundamental changes in order to be able to survive in its new environment, then it must be the environment that has fundamentally changed. And it has.

UNUSUAL BEHAVIOR
Macquarie University, Sydney, Australia, June 1998

There was a knock at the door. Lesley Hughes looked up from her desk to see a face appear in her office doorway, complete with bushy beard and a head of fair hair. It was an older colleague from the Institute for Biology. He asked Hughes if she would like to give the opening address at the annual meeting of the Society for Conservation Biology. The prestigious conference was due to be held outside of North America for the first time, at Macquarie University in Northern Sydney, where Lesley Hughes, then thirty-eight, worked as a lecturer and researcher.

Hughes was honored; it was the first time in her still-fledgling career that she would have the opportunity to speak in front of experts from across the world. She decided she would simply report on what she had been working on for the last couple of years: the effects that climate change might have on animal and plant species in the future. She had simulated how eucalyptus trees in Australia would react if climatic zones were to shift one day. But that was a long way off.

Still, she decided to look over some contemporary studies from other parts of the world. She wanted to be sure she was prepared. After a little research, Hughes stumbled across something particularly puzzling. A handful of studies from notable journals were reporting highly unusual behavior in a number of species. These were no longer predictions; they were observations.

Hughes read about ferns unfurling on alpine peaks in Europe, Mexican voles in the southwestern US leaving behind a great number of their habitats and colonizing sites farther north, and yellow fever mosquitoes spotted in Colombia at altitudes of over 7,000 feet.

The more she researched the issue, the more examples she encountered: Fish populations off the coast of California were replaced. Resident cold-loving species were dropping while warmth-loving species from southern California increased. In Great Britain, bird species were moving permanently up north; the same was happening in the US.

One study in *Nature* especially caught Hughes's eye: In the US, populations of *Euphydryas editha* had shifted northward by two degrees of latitude and climbed their way up mountains. This study stood out from the others in terms of its complexity, its volume of data, and the degree of care taken over the survey. It was published in 1996 and compiled by Camille Parmesan, a young biologist from the US. Parmesan had acted every bit the detective, canvasing museums in the western US for a whole year to use historical records to determine where the butterfly had lingered in the previous hundred years. She subsequently visited these sites herself to check whether there were still butterfly populations present. In four and a half years, Parmesan, at that time based at the University of Texas, observed over 150 sites in a strip along the West Coast from Mexico to Canada, including the San Ysidro Mountains next to the Mexican border. The result? Many butterfly populations in Mexico and the southern US had disappeared, while only very few had vanished in the northern US and Canada. Their center of distribution had shifted over 60 miles north and over 300 feet up. It was possible that these sensitive butterflies were serving as bioindicators for global warming, though this was as yet unproven at the time of publication. "Conclusive proof," Parmesan wrote, would require more such studies, taking in other species and other regions.

Lesley Hughes was astonished. It was precisely these studies that she now held in her hands. She was conscious that some of them displayed weak points or were merely snapshots. There could be all kinds of reasons behind a shift in habitat, rogue years where the weather was particularly good or poor. Animals and plants were constantly moving from place to place, if only randomly. It was also conceivable that people had driven these species to move, by appropriating areas of land or spreading environmental pollutants. Yet most of the studies, and the coincidence of the shifts they described, gave her pause. It all pointed to a pattern.

Eventually, Hughes gave herself permission to ask the big question: *Was the alarm bell already ringing? Has climate change already set the animals and plants of the world in motion?*

THE ALARM
Washington, DC, 1985

The idea had come to him in the shower. It was where he often had his best breakthroughs. Robert Peters, known to friends and colleagues as Rob, had just completed his studies in biology at Princeton and started his first job, at the Conservation Foundation, a conservation organization in the capital. Part of his job required him to write an article on ideal layouts for nature reserves. Across the country, people in environmental conservation were discussing which was better: one large nature reserve or multiple small ones.

It might sound like a minor issue, of interest only to a small group of experts, but it was anything but. The topic was a timely one. The habitats of animals and plants across the world were shrinking as humanity spread. More and more habitats were surrounded by towns and agricultural land or carved up by roads and canals. They were like islands in a vast ocean.

It was no accident that biologists like Peters turned their attention to findings in the field of island biogeography, a subject that deals with how island species develop and how they die out. Broadly speaking, they concluded that the farther islands are from the mainland, and the more isolated and smaller they are, the fewer species they harbor. Ultimately, interaction and exchange are crucial for biodiversity.

This could also be applied to fragmented mainland sites, such as patches of woodland or nature reserves. This information gave biologists the opportunity to calculate how quickly species in these locations would go extinct.

Peters immersed himself in the material before encountering another phenomenon by coincidence, which made isolated habitats an even bigger problem for species than they already were. What had long been considered protected space, could, in the long term, turn out to be a trap.

In the shower, Peters recalled an article he'd spotted in *Science*. The article by NASA scientists covered the possible ramifications of the greenhouse effect, a phenomenon that many had yet to speak of, and, if they had, they had only discussed it as a possibility in the distant future. The scientists spoke of shifting climatic zones transforming entire landscapes in North America and Central Asia into deserts and melting the West Antarctic Ice Sheet.[1] Peters pictured what would happen to living beings in nature reserves if, one day, the climatic zones were to shift away from the equator and toward Earth's poles. Vegetation zones would shift, too, and then—he was certain—the conditions that many species needed to survive would disappear overnight. Those who could not adapt would perish—unless they migrated. But where would they go? Would they leave the nature reserves? *Good God*, Peters thought, as the water pattered over his head. *It's a disaster waiting to happen!*

"A RIDICULOUS IDEA"

Peters visited the nearest library. He wanted to know what science had learned about this imminent threat in recent years.

He found nothing.

Then he spoke to conservationists.

No one knew anything about it. "It soon dawned on me that no one had ever considered it," Peters explains, with hindsight. "I felt like I'd found a twenty-dollar bill on the sidewalk, while everybody else had walked by without picking it up."

Something isn't right here, he thought. Peters knew he was onto something big, something that now makes him appear as a kind of a visionary; today, apocalyptic images of kangaroos hopping through charred Australian woodland are as much a familiar sight as bleached coral reefs or moose stumbling into Canadian supermarkets, covered in ticks.

Peters got in touch with Bob Jenkins, chief scientist at The Nature Conservancy, one of the biggest conservation organizations in the US, with headquarters in DC, close to the White House. Jenkins listened to what the young biologist had to say. Peters has never forgotten his reply.

"It's a ridiculous idea," he said. "Totally useless as far as conservation is concerned."

His reaction left a mark, even on someone as stubborn as Peters. Newly graduated and just at the beginning of his career, Peters was intimidated. He must have seemed crazy. But the matter was never far from his thoughts. He had fallen for it, hook, line, and sinker. And so, he asked a friend and colleague, the ecologist Joan Darling, to help him write a scientific paper. Unlike him, she knew what it took to get published. And the first thing they needed was more information. He found this by looking deep into the past, into the work of scientists with a predilection for scrabbling around in wet layers of sediment in lakes and wetlands. Here, paleobiologists were on the hunt for fossilized pollen. There are as many as a hundred thousand grains of pollen in one fifth of a teaspoon of lake sediment.[2] For paleontologists, this is true treasure, offering a glimpse far back into the history of life as we know it. Paleobiologists use special laser microscopes to observe pollen in 3D. Its shape reveals which genus it comes from, sometimes even which species. What's more, scientists can use its geological history to determine when and how many of a particular plant species have grown, because a new layer of sediment will settle on the lake bed every year. Since this layer is a different color in summer from what it is in winter, the sediment can be read in a similar way to the rings in a tree trunk. By observing these varves, as they are known, paleontologists can draw conclusions about early climatic fluctuations and how plants may have responded to them. How quickly did they spread over the course of millennia? How swiftly did their numbers dwindle?

This chronicle of Earth's history revealed to Peters a recurring, archaic phenomenon: Around every hundred thousand years, Earth enters an interglacial period that vitalizes all species of animals and plants and redistributes life across our plant. As if by silent agreement, on land and at sea, one species after another begins to migrate: insects and birds, amphibians and reptiles, mammals and fish, and even trees. They strive en masse to move closer to the poles, deeper into the oceans, higher up the mountains. They make use of the space left behind by retreating glaciers and masses of ice. If the climate changes again, turning cooler, the species move back. They follow an irresistible force, which attracts and repels them by turns. It's like a dance across the planet, one its inhabitants have performed dozens of times in the last 2.6 million years.

Darwin recorded this phenomenon over 150 years ago. In *On the Origin of Species*, he writes, "On the decline of the Glacial period, as both hemispheres gradually recovered their former temperatures, the northern temperate forms living on the lowlands under the equator, would have been driven to their former homes or have been destroyed, being replaced by the equatorial forms returning from the south."[3]

FROM NATURE RESERVES TO PRISONS

Something leaped out at Robert Peters as he conducted his research. The rate at which trees were migrating was considerably slower than the shift in the climatic zones. Many trees were being left behind, helpless. They were simply too slow.

The paleobiologists also had information about animals at their disposal. The chitin shells of bees were sometimes preserved in sediment for thousands of years, as were the bones of small mammals. Their remains revealed that animals were able to react much faster to climate fluctuations than plants, though this was of no use to them if the plants they needed to survive were missing from their new habitats. Peters grabbed some scissors, tape, and tweezers and set to creating a chart of his findings (computers were not yet widely available). The first image showed a reserve picked at random, with patches of cross-hatching designating a species' natural area of distribution. The second image showed the reserve still within the crosshatched area, but now surrounded by white patches— these were human settlements and cultivated areas.

In the third image, the reserve was now situated outside the crosshatched area; that is, the climate limits within which the species was able to survive. Peters concluded that "the consequences would be worst for those species which are restricted to certain areas, or which share the characteristics of species which are restricted to certain areas, that is to say, those with a limited range, those with small populations which are genetically isolated."

The nature reserves of today, he reasoned, would be the prisons of tomorrow.

Even if species had the opportunity to migrate in pursuit of their climate zones, their living situations would change fundamentally. A biological community does not migrate to a new place as a cohesive, settled unit as Darwin imagined. From pollen analysis, paleobiologists have learned that individual species, even individuals of specific populations, advance into new habitats at different speeds.[4] The result is that biological communities, as we know them today, split off into their constituent components. Some species go extinct, while others are able to survive in new places. The groups of species on Earth, Peters came to realize, were nothing but temporary communities of convenience—like a house share where the roommates are always changing.

However, this contradicted the prevailing theory of succession, which states that nature will ultimately find its way back to its original state following a disturbance (like storm damage) or human interference (like deforestation). "People had this deterministic belief that everything is more or less static," Peters explains, looking back. "What we consider to be stable communities are actually artifacts of earlier climatic events."

The makeup of species that live together is largely dictated by chance. "It was an exciting and terrifying thought," Peters says. "Everything could change."

And that's exactly what was about to happen: Earth's climate began to change again, the result of humans exploiting and burning much of the planet's fossilized energy resources, as well as razing countless of its forests to the ground.

Peters and Joan Darling submitted their article to *BioScience*, and an editor at the journal got in touch. He was interested, he said, but since the ideas behind it were so new, eleven peer reviewers would be called in to check the article.

Weeks later, they received an update: All eleven reviewers had rejected the article. They argued that it was simply too speculative. "Basically, no one believed it," Peters explains. But there were some exceptions. Tom Lovejoy, coiner of the term "biodiversity," was one. Another was the editor of the journal, who published the article despite the peer reviewers' opposition.

At least Peters can't say he didn't warn us. "The Greenhouse Effect and Nature Reserves; Global warming would diminish biological diversity by causing extinctions among reserve species" was published in *BioScience* in December 1985, taking pride of place between an article by Edward O. Wilson, the father of modern island biogeography, and one by Michael Soulé, the founder of conservation biology. Peters had linked both fields, developing his dystopian vision of the exodus of species, and combining it with recommendations on how to proceed: "If we are interested in preserving some remnants of the natural world for the year 2100 and beyond, we must begin now to incorporate information about global warming, as it becomes available, into the planning process."

By the year 2000 at the latest, NASA scientists explained, the climate change alarm bell would rise above the usual murmur of natural fluctuations in weather. Most species would then commence their long march across the globe: global warming made flesh. The most sensitive among them may well have begun a few years earlier.

AN AFFRONT
Sydney, July 1998

Lesley Hughes was nervous as she stepped up to the podium. The aging lecture hall at Macquarie University was packed to the rafters with seven hundred attendees from across the globe. Next to her on the stage sat professors from the University of Oxford, the Max Planck Institute for Terrestrial Microbiology (in Marburg, Germany), and Rutgers University, New Jersey. At least, thought Hughes, she would have the element of surprise on her side as she set her first slide on the overhead projector that afternoon on July 13, 1998.

She introduced her talk with a few general remarks, and then she dropped the bombshell: A whole raft of recently published analyses of long-term data sets suggested that some species were already reacting to climate anomalies. The migration of species had begun.

Hughes referenced a dozen such cases, including the migration of the little checkerspot in the western US, as reported by Camille Parmesan. For now, there were only examples of individual species shifting habitat. Yet it seemed inevitable that these isolated responses would grow into a vast cascade, increasingly influencing the composition and structure of biological communities.

Hughes invited her audience to take part in a thought experiment. They were to consider the species they worked with from a different perspective. "What happens when your species reacts to climate change?" she asked. "What would it mean for your research if your species started to move several hundreds of miles away?" What mattered now was to find out whether more animals and plants had already hit the road or might do so in the near future.

This request came as something of an affront, an attack on the conservationists' worldview. At the time, the prevailing notion was that the balance of species remains more or less stable, and this would persist for years to come. Each species had its ancestral territory. Nature reserves were considered to be the measure of all things.

"But we don't live in an equilibrium world," Hughes explained to her audience. "Eventually, even national parks won't be able to fulfill their role anymore, because most of the species will probably have to move out of the protected areas to stay within their climatically inhabitable zones."[5]

When Hughes ended her speech, there was friendly applause, and afterward people came and thanked her. "They were polite," Hughes recalls. "But I don't think most of them felt that what I'd said was groundbreaking."

THE DAWN OF A NEW FIELD OF RESEARCH

It can take a long time for new ideas to take root, particularly among conservationists. The term itself is just a few letters away from *conservative*. Yet with her overview of case studies, published in *Trends in Ecology and Evolution* in 2000,[6] Hughes had succeeded in heralding the dawn of a new field of research.

Since then, a whole host of biologists have analyzed shifts in the habitats of a diverse range of animals and plants. Initially, Hughes was often the only biologist at conferences concerned with the matter. These days, she finds herself attending conferences where *everyone* is talking about it.

The scientists rose to her challenge. In just a few years, a handful of cases became hundreds, and, two decades later, tens of thousands.[7] They all confirm that species across the world, from elephants to tiny diatoms in the ocean, are streaming toward the two poles. Land-dwelling organisms are retreating by an average of 10 miles per decade,[8] while ocean-dwelling ones are migrating by as much as 45 miles.[9] Life on Earth's surface is shifting away from the equator at a rate of 16 feet per day, northward in the northern hemisphere and southward in the southern hemisphere. In the oceans, this equates to a distance of 66 feet per day.

"What's surprising is that we're witnessing this across every continent and in every ocean," the butterfly researcher Camille Parmesan explained when I began looking into the matter, writing an article for *Natur* and *Bild der Wissenschaft*. "There is no region on Earth where this isn't happening, and there's no group of organisms that isn't affected."

How had I not heard about it? I was a little embarrassed. I had been working as a climate journalist since 2012, and yet it was only four years ago, and entirely by coincidence, that I came across the story of the mass migration of species, stumbling across a study that mentioned that cod had been reportedly migrating northward as the North Sea warmed. I had to read the sentence again. If cod were migrating to cooler waters, then surely other species of fish could well be doing the same? What if land animals were, too? What if all species were on the move?

I could only guess at the consequences it would have for humanity and nature alike, but I suspected they were vast. But I couldn't get more than a couple of select examples from the German conservationists I asked or from the articles I perused in journals. How could it be that a great redistribution of species was playing out across the globe—such as hadn't been seen for tens of thousands of years—and no one knew anything about it? Except, of course, the biologists who were researching it.

I decided to satisfy my curiosity and get to the bottom of it. I sifted through hundreds of scientific studies. It became something of an obsession and earned me the title of best customer at the photocopy shop on the corner. But almost every study would lead me to three new studies even more interesting than the last. I interviewed leading figures in the field and spoke to fishermen and foresters, and I traveled all the way to a mountain in the Peruvian tropics to learn more about a process that biologists call the "escalator to extinction."

I want the book you're holding in your hands to bring you along with me as I search for clues from the North Pole to the tropics, against the current of migrating species as I trace them to their source. I want to understand what will happen when this ancient mass phenomenon encounters modern civilization and turns life as we know it on its head. It already is.

Luckily, humanity has had time to adapt. Fifteen years before this "wave" swept across the Earth, Robert Peters had reported the phenomenon and warned of its consequences at countless conferences across the US and Europe. He even gave instructions for what we should do about it. The countries of the world should, he said, cut their carbon dioxide emissions in order to limit climate change as much as possible and soften the worst blows to the animal and plant worlds by creating new nature reserves and coming to the aid of migrating species, affording them more space or supporting them to settle in places where they are able to survive.

You're probably wondering if any of these measures were ever implemented. That's right: None were.[10]

And so, the largest field experiment in history, an ecological catastrophe in no uncertain terms, was allowed to run its course.

PART I

The Arctic: Predators without Prey

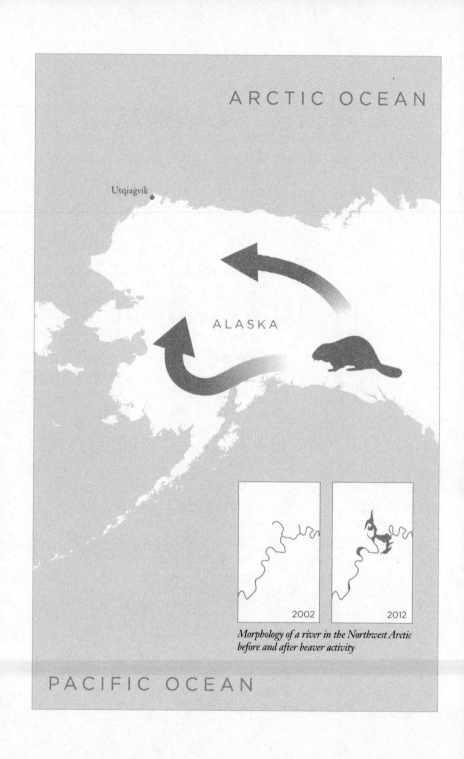

ARCTIC OCEAN

Utqiaġvik

ALASKA

2002 2012

*Morphology of a river in the Northwest Arctic
before and after beaver activity*

PACIFIC OCEAN

1
—
Hunters

Utqiaġvik, Alaska, 2015

The Indigenous peoples of the Arctic are the first to notice when the Earth seems thrown off balance and the rules that have held true for centuries no longer apply. They live in a region of the world that is heating up faster than the rest of the planet, and they are also more sensitive to changes in their environment. Since time immemorial, many fishermen, reindeer herders, and whalers have depended upon hunting to survive. This applies as much for the Sami in Scandinavia as it does for the Dolgans and Nenets people in Siberia, or the Yup'ik and Iñupiat peoples in Alaska.

Every few years, reports emerge of anomalies in the Far North. Take, for instance, a story from Utqiaġvik, the northernmost town in the US and the place where the Iñupiat people have made their living from hunting bowhead whales for generations. On January 29, 2015, Henry Huntington, a US polar explorer from Eagle River, southern Alaska, made a visit to them. He arrived on an Alaska Airlines flight—all roads in the four-thousand-strong town ultimately reach dead ends. Six days into his visit the sun decided at last to creep back over the horizon. It had been gone for two months, lying low during the polar night.

One frosty day, Huntington entered the Iñupiat Heritage Center, a long gray-white low-rise building, a kind of museum exhibiting the history of the Iñupiat and their relationship with nature. Inside, the sea ice expert spread a map of Alaska across a wooden table; the state's north coast forms a triangle with Utqiaġvik at its peak. The site had served as winter territory for the Iñupiat for centuries. In the nineteenth century, Europeans began to take an interest in this human outpost and built a whaling station.[1] As time passed, the settlement also attracted meteorologists and biologists, primarily for its location just 200 miles north of the polar circle.

Henry Huntington was similarly drawn to the place. He arrived for the first time in 1988, newly graduated from Princeton, hoping to record whale numbers. The job provided an opportunity to get to know the landscape of snow and ice he'd read about in books. In hindsight, he was almost ashamed to have known so little about the people and the nature of the region. Yet the whalers of Utqiaġvik welcomed him with open arms. They took him with them onto the ice where they camped out week after week, hunting bowhead whales. When they caught one of these giants, they would haul it up onto the ice to remove its innards and portion it up. Their whole culture revolved around this ancient act.

Huntington was captivated—and visited again and again. He visited in January 2015 and interviewed ten of the region's residents one after another, recording conversations with Willie Koonaloak, John Heffle, Ronaldo Uyeno, and others. They had plenty to say about the sea ice, where they hunted seals, walrus, and whales.

Sea ice is not static; it pulses like an organism in time with the seasons. In the summer, it retreats northward; in the winter, it descends once more. For many years, the sea in Utqiaġvik would typically begin to freeze in October. At the same time, there was also the uneven, thick multi-annual ice that crept out of the north as far as the coast and anchored itself there. This would provide the inhabitants of this remote area with a huge, walkable expanse from one day to the next.

All this is a thing of the past now. Sometimes seawater can still be seen sloshing along the coast as late as December, say the locals, some of them dressed in sports jackets and baseball caps. They practically never see the thick pack ice anymore. Without it, however, the newer ice lacks support; it floats and struggles to re-form over the winter. Year after year, hunters must travel farther out to sea in their boats in order to reach ice where they can land their catch. But again, in spring, the ice breaks up weeks earlier than it once did, making it dangerous for the Iñupiat to tread and travel on. Sometimes hunters struggle to find ice sheets thick enough to take the weight of landed bowhead whales.

Elsewhere in the world, other residents of the Arctic are increasingly struggling to recognize their home. It must seem utterly bizarre to the nomads and reindeer herders of northern Siberia. They report the ground giving way beneath their feet, whole hills collapsing and roads warping. Rivers now flow where they once felt hard ground beneath their boots; thousands of lakes have formed, such that the ground resembles little more than a net over the water. The reason for all this is hiding underground: a sixth of the earth on our planet is permanently frozen, a relic from the last ice age. As Earth warms, the ice in the ground thaws and transforms the whole landscape. The sea ice retreats. The Arctic is shrinking.

But this is just the beginning. Hesitant at first, the natural world is reacting to the reconfiguration of the landscape. It's on the move. Plants and animals are migrating en masse into the Arctic from more southerly regions and bringing with them new challenges to local ecosystems.

It's hardly surprising that researchers like Huntington are eager to learn from the Indigenous peoples of the Arctic. They need their help to understand how climate change is turning life on Earth upside down. To this end, they have equipped traditional hunters with digital cameras, encouraging them to photograph anything out of the ordinary—or they simply ask them. It began with reports of winters that were not as cold as they used to be. Then there was talk of shrubs slowly but steadily spreading northward as the years grew warmer. And then the animals followed.

In 1995, Huntington spoke to Iñupiat and Yup'ik elders in western Alaska about the disappearance of the beluga whale. At one point, he recalls, the conversation took a surprising turn, and suddenly all anyone could talk about was beavers.

Huntington must have looked puzzled because one of the old men smiled at him and asked, "Don't you see the connection?"

"I don't, to be honest."

"The local beaver populations are growing," the old man explained. "They dam the rivers. And that affects the fish. The fish migrate upstream to spawn, and later they return to the sea, where they live. The beluga whales wait for the fish where the river meets the sea."

Looking back, Huntington can spot early signs of the change. This shift is already throwing ecosystems in the Arctic into disarray and driving stoic communities like the inhabitants of Utqiaġvik to the verge of desperation.

A NEW MAMMAL ARRIVES
Fairbanks, Alaska, March 2017

Ken Tape stared at his computer screen, glancing from one satellite image to the other. Tape is an ecologist at the Geophysical Institute of the University of Alaska. He had heard rumors on one of his expeditions into the Arctic wilderness. A new mammal was said to have begun colonizing the Arctic. Looking at one shot, Tape frowned.

He had been working on the Arctic tundra for more than two decades. The landscape had hardly changed for many years, except for shrubs and small fields, which were spreading due to the warmer weather. But what Tape saw seemed like

The Canadian beaver is having a profound impact on the environment as it moves north into the Arctic, damming rivers and causing flooding that, in turn, melts permafrost.

someone had taken a hammer to the landscape. In place of the regular river courses visible in older shots, there was now a mosaic of lakes, sections of river, and wetlands—the kind he recognized from stretches of water that had been dammed by beavers. But the rodents had neglected to venture into the tundra thus far; there simply wasn't the food or the building material for their dams.

In the days and weeks that followed, Ken Tape and his colleagues examined high-resolution satellite images covering an area almost the size of New Jersey and found fifty-six beaver lakes that had not been there in 1999. "I've no doubt that we've got some industrious little furry engineers at work here and not some other natural process," Tape explains. From the distribution of the lakes, the scientists could even determine how quickly the Canadian beaver had spread along the coasts and rivers: an average of five miles per year. In twenty to forty years, *Castor canadensis* could have colonized the whole of Arctic Alaska, according to a study from 2018.[2]

In the intervening years, Tape has discovered thousands of lakes in the Alaskan tundra, all created by beavers. They fell trees, which are rooting farther and farther north; dam rivers; and flood whole stretches of land. Water transfers heat better than tundra vegetation, which means beavers are also causing the permafrost beneath and next to dammed waters to thaw. Experts speculate that these could attract salmon in the future. Another effect is already visible, and it's a particular cause for concern for Tape: Large quantities of greenhouse gases are being released from the ground into the atmosphere. "The great number of lakes are causing the permafrost to thaw dramatically."

The beavers are just the latest in a long line of newcomers. If trees and shrubs are spreading north and the ice is melting, conditions are also improving for other species: snow hares, white-tailed deer, and moose, which are colonizing the most remote corners of North America.[3,4] "It's following a pattern, one we've anticipated for some time," says Tape. "Boreal forest animals are moving into the Arctic."

The animals losing out to climate change are the long-established species: musk oxen, caribou, the Arctic fox. They are at a dead end[5]—and there's no way left to go. While shrubs and pine forests spread northward, these creatures' habitats are restricted in the north by the Arctic Ocean. It will be several decades yet before their habitats are irreversibly overgrown. But that doesn't mean they get a grace period—other animals are already migrating out of the south and toward the north, moving into their territory and presenting serious competition.

ARCTIC OCEAN

BARENTS SEA

KARA SEA

YAMAL PENINSULA

SIBERIA

RUSSIA

2

Hunted

FLIGHT OF THE ARCTIC FOX

For much of history, the Arctic fox was perfectly adapted to the chilly conditions in the Arctic. With its small ears and white winter pelt, it wastes little energy and can make itself invisible in a snowy landscape. But this is of no use to the fox if the snow thaws or its larger cousin, the red fox, moves in from the south. The red fox is benefiting from the warming of the Arctic: Less snow falls, and moose, reindeer, and people spread northward. It will find more carrion and more opportunities to shelter from the cold.

Population simulations in northern Scandinavia have demonstrated that only a relatively small number of these larger and heavier red foxes is required to push the Arctic fox farther north.[1] *Vulpes lagopus* had all but disappeared from the region: Just sixty adult animals remained in 2000.[2] The population was only able to recover to some degree following a breeding program, introduced by the Norwegian Environment Agency.[3]

Even in Russia, home to the most Arctic foxes in the world (as many as eight hundred thousand, according to estimates), biologists were for a long time unable to observe any takeover of this kind. That is, until July 22, 2007, when Anna Rodnikova, a zoologist from Moscow State University, visited an Arctic fox den on the Yamal Peninsula in northwestern Siberia. She had been observing a vixen and her pups for a week.

That day, however, the situation changed. Shortly after 6:00 PM, Rodnikova spotted a red fox passing the den, just over 100 yards away.[4] She noticed that the red fox would approach the den slowly and stiffly. He kept pausing to pant.

After half an hour had passed, the Arctic fox returned home from her foray. She approached upwind of the den, slowing her pace and creeping closer—she had clearly smelled her adversary. On the hill, a stone's throw from Rodnikova's position, she lay down on the ground and remained there for twenty minutes, keeping her eye fixed on the entrance to the den. As the red fox emerged, the Arctic fox barked at him from a safe distance. When that did nothing, she turned on her heel and ran, leaving her pups behind. "Despite the red fox being in a bad state, the Arctic fox made no attempt to tackle the intruder," Rodnikova wrote in *Polar Biology* in 2011.

THE NEXT ICE AGE IS CANCELED

It's a tragic twist of fate for the animals of the Arctic. Normally, the age of the Arctic dweller would be dawning now. Following the intermezzo of the Holocene—an interglacial period lasting ten thousand years, which has provided human beings with a stable climate in which to develop from hunter-gatherers into farmers—Earth should have plunged into its next ice age long ago.

If this were to occur, Arctic foxes would be able to capture ever larger regions of Central Europe due to the spread of the cold, just as they did in the last ice age, even colonizing southern France, without any notable competition.

Sixty-five hundred years ago, the Earth was oriented such that summer sun was at its lowest level in the northern latitudes—a sign, in fact, that a cold spell was about to begin. And it would have if it hadn't been for carbon dioxide, which began collecting in the atmosphere. Climate scientists are still debating why. Some view changes in ocean currents as the cause, while others believe our ancestors were the source of the issue, clearing the forests to grow crops and graze cattle.[5]

With the advent of industrialization, the pendulum swung definitively in one direction. The world is warming considerably—predicted warming may even proceed with greater force than has been seen for two million years.[6] The Arctic Ocean could be completely free of ice in summer as soon as the middle of this century. The ice age is canceled.

Given the speed at which this change is occurring, most Arctic species have no chance of evolving to adapt to climate change.[7]

They have a greater chance of success if they breed earlier or later, depending on their habitats, or decamp to summer or winter pastures. But those who can't reset their body clocks will have no choice but to migrate to the Far North.

"As the warmth returned, the Arctic forms would retreat northward, closely followed up in their retreat by the productions of the more temperate regions," Charles Darwin once wrote.[8]

Arctic foxes, musk oxen, and caribou still have large areas in which they can comfortably live. Yet every degree farther north they retreat, the Earth's geophysics outwits them: The closer they get to the North Pole, the more the inhabitable territory shrinks. Earth is an ellipsoid, after all, much like a medicine ball. Let's imagine a ring around the equator, moving north and then contracting, tighter and tighter. The habitat available to animals and plants is constantly shrinking. Scientists call this the "polar squeeze." Some species, like seals or polar bears, have literally made their way to the ends of the Earth.

It seems like a paradox. The diversity of Arctic species is growing thanks to migrants from the South. At the same time, however, the Arctic's contribution to biodiversity across the world will drop, because polar species die out if they cannot find a way to survive in the era of climate change.

THE PIZZLY: EVOLUTIONARY ACCIDENT, OR NEW TREND?

Banks Island, Canada, April 16, 2006

It was a bracingly cold day and a sixty-six-year-old man was stomping through the snowy landscape of the Canadian Arctic, dressed in a white winter coat, a ski mask, and goggles. Jim Martell, owner of a telephone company in Idaho, had received a sign from his guide: A polar bear was in sight, close enough to attempt a shot. Martell had paid $50,000 for his license to shoot polar bears, a practice that outraged conservationists. Martell didn't care. He took aim, fixed his eye on its creamy white coat, and pulled the trigger. The bear slumped to the ground.

When Martell bent over the animal, he noticed it was quite small for a polar bear, with a hump and long, brown claws. He surveyed the grubby-looking pelt and the dark flecks around its eyes and snout. Had he perhaps shot a grizzly instead? If he had, he risked up to a year in jail; this subspecies of bear was protected in Canada.

A DNA comparison revealed that it was neither polar bear nor grizzly, but a mixture of the two: a hybrid, born to a polar bear mother and a grizzly father.[9] The press heralded it as a "pizzly" or "grolar." Scientists considered it a rare anomaly; some saw it as an evolutionary accident.

But then other hybrid bears emerged. One of these even proved to be the progeny of another pizzly, in gene analysis.[10] It had already been established that polar bears and grizzly bears were able to mate, since the two species were closely related. However, they had mainly stayed out of each other's way in the past, which was why the discovery of several hybrid bears stunned biologists, confused taxonomists, and presented conservationists with a challenge.

Climate change provides one explanation for this phenomenon. Thanks to global warming, the habitats of *Ursus maritimus* and *Ursus arctos* are increasingly overlapping, offering more opportunities for unusual pairings. Male grizzly bears are straying into the northernmost regions of North America

The habitats of grizzly bears (top) and polar bears (bottom) were once largely separate, but climate change is bringing them together, setting the stage for hybrid offspring between the two: the so-called pizzly or grolar.

because they are less vulnerable to hunting there and, thanks to global warming, are able to spread to areas with climates that would previously have been inhospitable; the polar bear's habitat is shrinking, because the sea ice—where it hunts seals—is retreating.[11] When it finds itself faced with nothing but open sea, it is obliged to flee on land to places where it may meet not just humans, but grizzlies, too.

Some Arctic researchers hypothesize that grizzly bears will, sooner or later, capture the High Arctic tundra, while polar bears gradually disappear. Prognoses estimate their populations dwindling by half by the middle of this century. In the future, these Arctic residents, once the absolute rulers of an enormous kingdom, will have to seek refuge to survive. Eventually, the High Arctic in Canada and Northern Greenland will be the only retreat for polar bears. If they go extinct completely, their genes, at least, will live on in the grizzly bear population—much like the genes of the Neanderthals, a small portion of which live on in modern-day humans.

For the Arctic fox, biologists assume, the only remaining havens are the islands in the Arctic Ocean.[12] There at least, climate change will suit them for once: As the sea ice disappears, the links between the islands and the mainland are broken. This could well save the Arctic fox from the red fox—if it's able to hide out there for thousands of years, until the next ice age dawns.[13, 14]

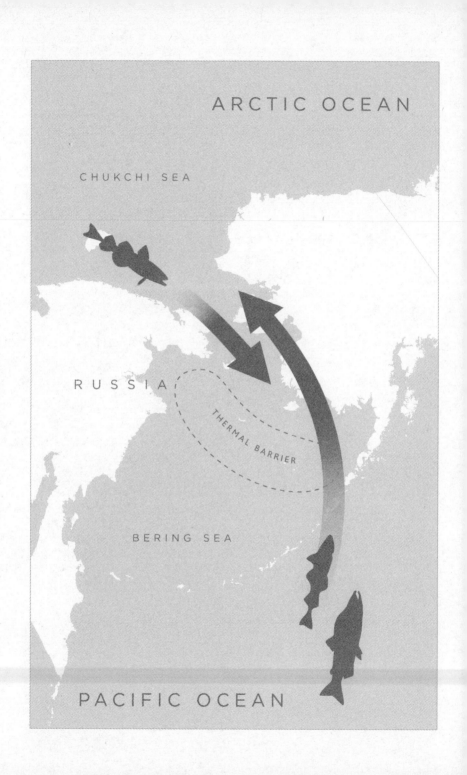

ARCTIC OCEAN

CHUKCHI SEA

RUSSIA

THERMAL BARRIER

BERING SEA

PACIFIC OCEAN

A Change of Regime in the Ocean

THE THERMAL BARRIER FALLS

As a struggle for survival emerges in the world's most northerly regions, the process is already well underway in the Arctic Ocean. Water-dwelling organisms respond much more rapidly to warming. Many fish and whales change location daily to remain within their optimal temperature zones. They do this by sinking to greater depths or migrating northward. If a sea like the Bering Sea warms severely, bearing less ice, its inhabitants may be replaced in a matter of months.

Bering Strait, 2017

Ten thousand years ago, Asia and America were still connected by a narrow land bridge. When the last ice age ended, vast bodies of ice melted, sea levels rose, and the Bering bridge was submerged. Today, a gap fifty miles wide yawns between Siberia and Alaska; it is through the eye of this needle that water from the Pacific rushes into the Arctic Ocean, and vice versa. Yet exchanges between boreal and Arctic inhabitants of both oceans remain restricted to a minimum. This is due to a barrier, invisible to the naked eye. On the ocean floor south of the Bering Strait, there sits a layer of water 100 feet thick, cooler than the surrounding water by 2°C (3.6°F) at most, fed constantly by melting sea ice. This reservoir inhibits free flow between the two systems like a huge thermal wall.

Then came 2017. The oceans had been accumulating heat for years. As the sea ice melted, the areas of dark, open sea grew, no longer reflecting the sun's rays but absorbing them instead. It was a self-perpetuating cycle. The sea ice retreated farther north from one summer to the next, withdrawing from the Bering Sea, until it disappeared behind the shelf edge in the Chukchi Sea. Strong winds also caused more warm water to stream through the Bering Strait into the marginal sea of the Arctic Ocean; the once cold-water basin spent much of 2018 and 2019 above freezing. Henry Huntington, heading up a group of experts, described the effects as "shocking."[1]

Now that there was no longer a thermal barrier in front of the Bering Strait, there was a sudden rush of comings and goings underwater: Polar cod paid a visit to the Bering Sea from the north, which marine biologists dismissed as an anomaly, blaming it on peculiar conditions that year. However, Alaska pollock, a warmth-loving fish species, made their way through the strait in the opposite direction, northward. So too did pink salmon, streaming along the coast in great numbers as far as Utqiaġvik, threatening populations of whitefish, a primary source of food for the whalers on the ice. "We didn't see it coming," says Huntington. "We hadn't expected these conditions to present themselves for another few decades."

This wasn't the first "regime change" of its kind, however. In 2015, Norwegian scientists noted that a similarly radical change had taken place in the Barents Sea. For ten years, they set off each autumn with their research vessels from four hundred stations in the marginal Arctic Ocean in northern Europe, cast their trawler nets, and took samples. Then they classified the fish, counted them, and weighed them. The results demonstrated that while predatory fish like cod or haddock were shifting their centers of population as much as 100 miles northward from the boreal—i.e., cold-temperate—zone in the Atlantic, and moving into Arctic waters, they were forcing local fish species like Greenland halibut, deep-water redfish, and polar cod to retreat into more northerly waters. Project head Maria Fossheim, from the Institute for Marine Research in Tromsø, described this phenomenon as the "borealization" of fish populations in the Arctic.

However, what Henry Huntington observed in the Bering Sea off Alaska was not limited to fish. He collated the curiosities that specialists had observed: sea birds dying in droves along the coasts of Alaska, and seals, too; whales suddenly opting for routes different from those they had swum for centuries. "Everyone was reporting unusual events," Huntington explains. "It gave us the impression that these were something more than isolated phenomena."

The change would have immediate consequences for traditional hunters in Alaska. "Eskimos never stay hidden," went the saying overheard by Huntington in a town in the west of the state: It refers to snowmobiles but can be interpreted as meaning, essentially, "If you've got a problem, see how you can fix it." In the past, this attitude helped hunters to survive the harsh conditions. "But everything has its limits," Huntington says. "The best attitude in the world couldn't have prevented what happened in Utqiaġvik in the autumn of 2019."

Utqiaġvik, September 2019

Whaling season began with great anticipation. What else was to be expected? Every autumn, seventeen thousand bowhead whales traveled westward from Canada, passing Utqiaġvik, steering clear of Alaska, and turning into the Bering Strait before migrating along the Siberian coast to the south, where there was plenty of krill and other small crustaceans to eat at that time of year. It was as dependable as the imminent onset of the polar night. As they journeyed past the coastal town, the ice from the north would practically steer the huge sea mammals straight into the hunters' laps. All they had to do was wait, just like the generations before them. The iconic whale bone arch that stood on the coast, each bone several yards long, was testament to the whalers' prowess.

These days, the whales are protected, and can live up to two hundred years, but the residents of Utqiaġvik are permitted to catch twenty-five of them each year for their own supply, as are other Indigenous communities along the north coast of Alaska. The whales feed the town in winter. Not only that, but the nutty whale meat has an almost religious significance. "It has been at the heart of their culture for generations," Huntington explains.

Everything is geared toward whaling. In the summer, when the ice retreats, the men of Utqiaġvik hunt bearded seals and use the sealskin to cover the boats they take on whale hunts. In summer and autumn, they hunt caribou; its flesh nourishes them on the ice and its pelt keeps them warm. If the whale hunters are successful, they throw a great feast and the whole community dines on whale meat. "They have to spend most of the year getting everything ready," Huntington explains. "It's an enormous undertaking for the whalers, but it's also the social glue that holds the community together. If you take the whales away from them, you take away what drives them through the year."

An exceptionally hot summer had just come to an end, with temperatures reaching 90°F. July in Alaska was never this hot. Salmon perished in the rivers, and even the forests were burning. The ice in the sea outside Utqiaġvik only began to creep hesitantly toward the coast in autumn. But the residents of the town had no doubt that the whales would pass by once again when the time came. The great sea mammals were reliable, even in the years that followed the curious changes in the Bering Sea in 2012.

The whalers set out on their boats under cover of dark. The men would spend up to fourteen hours at sea at a time, looking for whales.[2]

But they didn't find a single one.

Expectations grew with each week that passed, but the whales never appeared.

Once four weeks had passed, the first of the men began to pack their boats away. They could not afford to spend any more money on gas or food for the whaling teams. The emotional pressure was growing, too. The whalers were away from their families for weeks. It was a sacrifice they were willing to make as long as they eventually returned with a good catch, one that could feed the community for weeks and months. But returning home without a single whale after weeks away was humiliating.

September passed, then October, all without a single whale.

Now the hunters were running out of time. The sun's light lessened with every passing day; polar night was approaching. It would soon be too dangerous to go out to sea, even more than it already was. Just that past fall, two whalers had drowned when their boat capsized in stormy seas.[3]

Only the boldest still sought their fortune on the waves. Sometimes they would sail out more than fifty miles to seek the whales, farther than ever before—but without success.

Catholics gathered in their church houses to pray, as did Presbyterians and Adventists. "When you begin to lose the light and hope that once lit bright inside, you realize that maybe that's when the miracles will happen," one townswoman explained to a local reporter.[4]

ON THE TRAIL OF A BORN SURVIVOR

At the same time, a state-of-the-art research ship was crossing on the other side of the seventy-mile zone. It had sailed through the Bering Strait and circumnavigated the west coast of Alaska. It was not permitted to dock in Utqiaġvik; the crew aboard the *Sikuliaq* were not welcome. "The whole community was in an uproar," explains Hauke Flores, one of the expedition members from the Alfred Wegener Institute for Polar and Marine Research in Bremerhaven.

A sea-ice biologist by trade, Flores was also interested in the radical ecological changes taking place in the region. More precisely, he was interested in a small fish that plays a key role in the Arctic food web and serves as an indicator species for the state of the whole ecosystem: the polar cod. Evolution has equipped the polar cod, a born survivor, with three dorsal and two anal fins, making it perfectly adapted for life under the ice. Scientists believe that young cod between one and two years of age allow themselves to be carried from the spawning grounds around the coasts of Siberia and Alaska and into the central Arctic, protected by the pack ice. Beneath the ice, the young fish are safe from fulmars and ringed seals. They feed on copepods and shrimp, which in turn feed on ice algae.

The polar cod can survive at temperatures close to freezing thanks to an antifreeze protein in its blood. "Animals that manage not to freeze in this region hardly use any energy at all," Flores explains. "All metabolic processes occur extremely slowly."

Yet this evolutionary advantage is lost if the sea ice disappears. Ironically, it is only now, as the polar cod's world is falling apart, that scientists have had the opportunity to understand how it lives. "At the very least, I want to record a realistic snapshot of the old Arctic for posterity," Flores says.

It's one of the reasons he traveled to the Beaufort Sea. But there was no trace of the polar cod outside Utqiaġvik, despite its having once been a regular guest to these waters. Instead, the researchers found species of fish in their nets that didn't belong there at all. "The whole ecosystem had clearly been turned upside down," Flores explains.

The ice sheet had reformed, but it was thin enough for Flores to reach down and grasp the bottom. There was no sign of the irregularly shaped multiannual ice. The *Sikuliaq*—the Inuit name for young sea ice—pushed on across the edge of the shelf and out into the sea. The crew cast a special trawl net that stretched out under the ice and caught young polar cod way out

The polar cod, an indicator species for the state of the ecosystem, produces a natural antifreeze protein that enables survival at temperatures near freezing—an evolutionary advantage that's lost if sea ice disappears.

in the ocean. "In the past, they didn't have to swim out into deeper waters to get under the ice," says Flores. "In previous years, the ice would reach the shallow waters in autumn."

The spread of the Arctic sea ice in summer has almost halved since 1980. Consequently, young polar cod have had to cover greater distances to migrate back from their icy refuges to the spawning grounds on the coast. In theory, polar cod can also survive in the open ocean: Laboratory experiments demonstrate that they can cope with rising temperatures to a certain degree. However, they soon find that they are not alone. Species like Atlantic cod press in from the south and are better able to assert themselves in the new conditions.

In part, it has something to do with the new menu: If the Arctic sea ice retreats, copepods—a young polar cod's main prey—find little or no ice algae to graze on the underside of the ice. Atlantic and Pacific species of copepods emerge in their place, much smaller and less fatty. The polar cod has to gobble more of these in order to eat its fill. But there is other competition for this meager snack, from fish species from the south like caplin and herring. And they also attract Atlantic cod, which pursues its prey for hundreds of miles and has been known to kill off polar cod, too.

Ultimately, there's nothing left for the polar cod to do but swim out deeper into the Arctic Ocean. But in one respect, they still have luck on their side. A fishing ban is in place—for now.[5] By contrast, other species of fish can survive only in the relatively shallow shelf seas close to the coast. They don't have the option of retreating toward the North Pole. "There is a whole range of species that will not withstand the change," says Flores.

First ice algae, then copepods, then polar cod: Change is cascading all the way up the Arctic food chain. The same is true in Alaska, and this began in the winter of 2017. The ice spread only thinly across the sea, leaving ice algal blooms weakened and masses of crustaceans starving. Fish like the polar cod could not find enough to eat; this had far-reaching ramifications because these industrious little migrants serve as an energy supplier between lower and higher organisms in Arctic waters. Seals and seabirds are specially adapted to eating polar cod. This may explain why hundreds of corpses of seals and guillemots have washed up on the coasts of Alaska in the years since, including outside Utqiaġvik.[6] "The change in the polar food web is clearly very much underway," says Flores.

And it continues up the food chain. In August 2019, more than two hundred gray whales washed up on the west coast of Alaska. They had starved to death. It appears that they could not find enough to eat in their summer feeding grounds in the Chukchi and Beaufort Seas.

But bowhead whales had seemed to have escaped the changes unscathed. They appeared outside Utqiaġvik every autumn without fail. Until 2019.

ARCTIC OCEAN

CHUKCHI SEA

Utqiaġvik

BEAUFORT
SEA

ALASKA

CANADA

GULF
OF ALASKA

4

Where Are the Whales?

Northeast coast of Alaska, October 29, 2019

Megan Ferguson wanted to try her luck one last time. At 9:30 one lightly overcast morning, Ferguson, a biologist from the National Oceanic and Atmospheric Administration (NOAA), boarded a small aircraft ready to fly over the ocean off the northwest coast of Alaska to search for bowhead whales. It was no easy undertaking. Sometimes clouds of mist would obscure the view, and the dazzling sunlight was blinding.

The whales had made their way from the east, along the coast of Alaska, so regularly in previous years that, on her test flights in September, Ferguson had enjoyed guessing the depths at which she would encounter the huge sea creatures each time. On this fall day, however, everything was different. At first, the whales did not appear at all, and later they emerged much farther north than their usual migration route. It was an ecological mystery.

Why are the whales so far from the coast? Ferguson wondered.

To this day, she doesn't have a clear answer. It's possible that something actively drove the whales away from the coastline—a fishing fleet, for instance. But there had been no unusual increase in the number of trawlers.

The situation is different when predators, such as orca, migrate up out of the south. These black-and-white mammals are also known as killer whales for their ingenuity and brutality when hunting seals.

Growing up to thirty feet long, orca are not whales in the truest sense; they are actually the largest species of dolphin. They prey on anything they can get between their interlocking teeth: fish, seals, whales, narwhals, and bowhead whales, too. The bowhead whales had previously sheltered amid the sea ice, as predators avoid it for its sharp edges. It's in open water where the feeding frenzy really begins.

For some years, scientists observed orca spreading northward in Hudson Bay,[1] but also in the western Beaufort Sea outside Utqiaġvik. They also registered an increase in attacks on bowhead whales in the region. In August 2019 alone, they found five cadavers bearing bite marks from killer whales: semicircular wounds on the skin, one animal disemboweled, tongues missing. Only one such case had been documented so far north in the ten years prior to this. It presents yet more competition for the Iñupiat.

That fall day in 2019, Ferguson stared at the dark blue of the ocean's surface. She was searching for the rest of the population; the team had by no means succeeded in locating all the whales two months before. As the plane flew across the sea north of the Bay of Prudhoe in the northeast of Alaska, Ferguson saw them coming.

Huge bodies broke the water's surface, the pitch-black of the backs of the whales was overlaid with white wounds from the sharp corners of the ice. With each breath, they blew out powerful fountains of water. Thirty bowhead whales, eight of them calves. They were splashing around, exactly where Ferguson had been waiting for them two months before. "It indicated that part of the population had delayed their migration," she says. "Perhaps the feeding conditions in the eastern Beaufort Sea were so good that they remained in the area."

Either way, they were on the move at last.

Utqiaġvik, November 16, 2019

Rumors spread across town at lightning speed: A whale had been caught off the coast.

The news was barely circulating that morning before the crew's wives began preparing their homes and their kitchens. Then, they ventured out into the darkness, got in their cars, and a caravan of red lights snaked its way through the hilly landscape to the coast, where the Iñupiat ancestors had hunted whale over six hundred years before.[2] They waited there for the whale. "People were hugging each other, crying, and screaming with joy," one of the townswomen told a local newspaper.[3]

Quilliuq Pebley and his crew had gone out for another try on that still, frosty day, when the sun hardly shone. The remaining boats coordinated over walkie-talkies. When Pebley spotted the bowhead whale, he gave the others the GPS coordinates and all the boats swarmed together to shoot the nearly twenty-six-foot-long whale with harpoons. They dragged it onto the ice, where they used butcher's knives to remove its head, and hauled the behemoth back to shore.

"So, to see the whale pulled up, you exhaled, like I've been holding breath for a long time," one of the town residents told Alaskan reporter Shady Grove Oliver. It had been two months of praying and steering their boats back out to sea, and they had used almost 900 gallons of gas. "I need a new word for how it feels because it is a mix of emotions: safe, complete, relief."

According to the report, a flag was raised on the captain's roof; a quarter of the townspeople gathered beneath it, over a thousand men and women. Each of them wanted a piece of the whale.

Henry Huntington once took part in one of these ceremonies himself. Like everyone else, he arrived one afternoon carrying a dish, cutlery, and a cup and sat down in the circle which had formed around the feast. There was *maktaaq*, the skin of the whale and the rind underneath, which is cut out in blocks.

"If the meat is from a young whale, it can be very tender," Huntington explains. "Old whale meat is quite tough."

Early in the evening, guests at the feast took the seal skin from one of the whaling boats, stood apart pulling it taut, and then tugged at it from all sides, bouncing one guest after another on the skin like a trampoline.

Yet as great as their joy was, the year had left its mark on the residents of the town. They had been able to depend on the whales for centuries, but now that was over.

Elsewhere, too, species on which the Indigenous peoples of the Arctic depend, around which their cultures are built, have evaded them. The Sami in the north of Finland have fished for salmon for millennia—it is what has ensured their survival to this day. "If the salmon go, we will no longer be people," one Sami saying claims. Yet this deep spiritual connection is at risk of being broken, in part because rising temperatures are causing another species of fish—pike—to migrate farther upriver and expel the salmon from their usual habitats.[4] As with the pike, the same is true for many other species on the move in other parts of the world. "They don't know how to live with these new species and their spirits, because they have no songs for them, no poems, and no art," Gretta Pecl tells me. She is the director of the Centre for Marine Socioecology at the University of Tasmania, which has worked with Indigenous groups across the world.

In northern Siberia, Europe, and America, the cultures and economies of many Indigenous communities depend on reindeer and caribou; ceremonies and dances are performed for them, their meat is eaten and sold, and their pelts are used for clothing, shoes, and tents—as in Utqiaġvik.[5] But even these great herbivores are moving north as the climate changes, changing their migration routes because the permafrost is melting, or facing competition from moose from the South,[6] which also carry pathogens against which Arctic species have no resistance.[7]

There's still hope for the people of the Far North. The Iñupiat, for instance, have become highly skilled at adapting to their surroundings. If one species disappears, even one that is deeply significant in their culture, they will simply choose to hunt another one and develop a new connection to that species.

The biggest problem for traditional hunters might not be bowhead whales or the salmon that fail to appear, nor the orca or pike advancing from the South, but another dominant species that is spreading north and forcing its own rules on Arctic communities: modern man. The residents of Utqiaġvik are only allowed to catch a small contingent of bowhead whales— they have no fishing rights to the minke whale, for instance, which surfaced right in front of the whalers' boats in the fall of 2019. Huntington is critical: "A man-made system has come into play here which, for no good reason, prevents any adaptation to climate change."

The local fishermen and whalers in the Arctic may not be alone for much longer. The Bering Sea in western Alaska is home to some of the most lucrative fishing grounds in the world. In 2017, Alaskan fishermen caught 1.3 billion dollars' worth of Alaska pollock in local waters. If fish species are migrating toward the North Pole, then it's only a matter of time before commercial fishing fleets follow—fishing ban or no.[8] Polar scientists have already found evidence of trawler nets at the edges of the ice.

This would change the nature of fishing in the Far North: The local subsistence economy would be replaced by mass industry. The figures illustrate best what this truly means: In over fifty years, Indigenous fishermen in the Arctic have caught fewer tons of fish than fleets in the Northeast Atlantic currently catch in a single year, and that's just from herring stocks.[9]

In the future, trawlers outside settlements like Utqiaġvik could contest the rights of the Iñupiat to the fish, and possibly drive the whales away, too. Their concerns seem justified. When it comes to fish, even wealthy nations and "civilized" democracies have no sense of humor. Especially when fish don't recognize national borders.

PART II

New Residents in the Temperate Zone

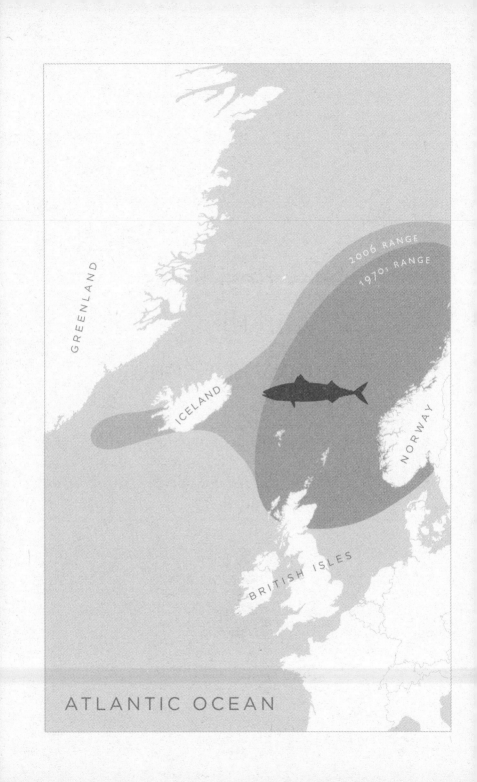

2006 RANGE

1970s RANGE

GREENLAND

ICELAND

NORWAY

BRITISH ISLES

ATLANTIC OCEAN

5
———
The Bread-and-Butter
Species Swim Away

In 2007, a school of mackerel set off from the edge of the North Sea and caused a trading war and a political crisis in Europe, the consequences of which can be felt to this day.

At the advent of each new year, millions and millions of mackerel gather off the coast of Europe to spawn on the continental slopes from Gibraltar to the Bay of Biscay, all the way up to Scotland. As spring draws to a close, they join together to search for feeding grounds between the North Sea and the Norwegian Sea.

In 2007, however, something happened that no one had anticipated: The Northeast Atlantic mackerel stocks expanded far beyond their usual grounds, advancing close to the coast of Iceland. For the people of Iceland, it must have been akin to a Biblical event. Witnesses report seeing a dark shadow looming in the sea; there was so much movement beneath the surface that the water appeared to be boiling. "In a few days, there were no other fish left," explains Ragnar Árnason, professor of economics at the University of Iceland in Reykjavik.

Before 2007, Icelanders had mainly fished cod, rose fish, and haddock; they had only ever caught mackerel in small numbers. But now, vast amounts of this oily delicacy had appeared on their doorstep. The balance that existed among the fishing nations of Europe had shifted almost overnight.

For years, Europeans had passed over the oily fish at the supermarket counter; in restaurants, they preferred to order salmon or trout. Mackerel simply wasn't in. The early '90s, however, saw a rediscovery of mackerel and its succulent, flavorsome meat, primarily among eco-conscious shoppers. Mackerel contains a high proportion of omega-3 fatty acids and, since it had been largely ignored for many years, fish stocks were in good condition.[1]

Iceland could hardly believe its luck, but the EU and Norway were less enthused. Until then, they had managed the stocks—and made a lot of money from them. The mackerel made over 220 million euros for the British fishing industry alone.[2] The Brits shared fishing rights with the other nations in whose waters the mackerel swam, such that total stocks had somehow remained stable. They had also come to an agreement with Iceland, determining how much each nation could catch; however, in view of the new situation, the island nation no longer considered itself beholden to this and took advantage of the stocks in its waters.[3] It was only a matter of time before the conflict escalated.

COLD-WATER SPECIES RETREAT NORTH

Strictly speaking, it's nothing special for mackerel to move to different waters. They are schooling fish, with vertical stripes on their backs: the true torpedoes of the sea. They have to be moving constantly because they do not have swim bladders, which fish usually use to float.

To this day, however, marine biologists are still puzzled as to what exactly led the Northeast Atlantic mackerel stocks to expand so far northwest. There are two possible explanations. The first speculates that mackerel stocks in the North Sea could have grown so huge that a number of them chose to seek out new feeding grounds. The second points to the fact that the waters around Iceland have warmed by 1–2°C (1.8–3.6°F) in the last twenty years.[4] "The two theories are not necessarily mutually exclusive," explains Ragnar Árnason. The stocks may have grown as a result of climate change, then spread.

Alternatively, they may have grown and then been able to spread because the waters in the North have warmed.

It would certainly fit the trend that is being observed in the North Sea: Fish species that were once observed off the coast of Germany in vast numbers are gradually dwindling, including herring in the Baltic Sea[5] and cod in the North Sea. The North Sea has warmed by 1.7°C (3.06°F) on average since the early '80s.[6] And this influences cold-loving fish species if, like cod, they find themselves living for decades at the upper limits of their temperature tolerance: If the oceans warm, so too do the fish. Their body temperature matches the temperature of the surrounding water at a couple of seconds' delay. To remain in their ideal temperature region, they must constantly react to fluctuations in temperature, dodging up and down, this way and that, to elude unwanted shifts. When the ocean warms permanently, this boosts a fish's metabolism; its heart beats faster and pumps more oxygen through its blood vessels. But since the oceans contain less oxygen the warmer they become, fish are subjected to twice the stress. "It's like running a race at high altitude," says physiological ecologist Felix Mark from the Alfred Wegener Institute for Polar and Marine Research. "And afterward, your muscles would never stop aching."

This has consequences for females, which reach their limits quicker when packed full of eggs. If the spawning grounds have warmed to an excessive degree, a fish will no longer feel content to lay her eggs there and will absorb them instead. The chances of her offspring—and perhaps she herself—dying are simply too high.

However, it is fish larvae that exhibit the greatest sensitivity to changes. They have to expend all their energy in order to grow out of this most dangerous of phases as quickly as possible.[7] Any additional exertion can mean the difference between life and death, which is why our warming seas pose such a threat. And if their main food source goes missing—because the zooplankton no longer keep to their annual rhythms[8] or

gather hundreds of miles further north,[9] as is currently happening in the North and Baltic Seas—it will be impossible for
them to survive. Whole populations of cold-water species in
the south are breaking away, only to reappear in the north.[10]
According to research into trawler fishing, species of fish such
as cod and mackerel in the North Sea have shifted their centers
of distribution by as much as 250 miles toward the North Pole
and as much as 12 feet deeper per decade.[11]

This presents a serious challenge for the fishing industry.
Fishermen are often conservative and fish in the regions they
have always fished. Only very gradually do they adapt their
operations to new conditions—or they go under. Dozens of
family businesses in Germany went bust because herring and
cod were frequenting local waters less and less. The few fishermen who were still clinging to cod had to spend millions of
euros equipping their boats for the high seas and then ventured
into ever deeper waters, heading North to pursue the fish that
had once been their bread and butter.

Small businesses on the East Coast of the US fared similarly.[12]
Only the large deep-sea fishing outfits were able to stay afloat: If
they specialized in just a few species, their tugboats generally
migrated north.[13] Take, for instance, a whole fleet of fishing boats
from North Carolina and Virginia, which spent many years fishing close to their home harbors and are now employed 500 miles
farther north off the coast of New Jersey. But this only works if
the fish keep within the borders of the specific territory.

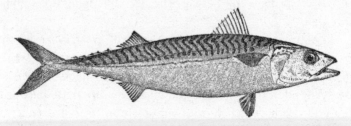

*A warmer ocean is a double stressor for the mackerel: Oceans contain less oxygen the
hotter they become, and yet they raise the Atlantic mackerel's need for it by boosting the
mackerel's metabolism, its heart beating faster and pumping more oxygen.*

"YOU'RE STEALING OUR FISH!"

When the first mackerel appeared in their waters shortly after the turn of the millennium, Icelanders didn't know what to do with them. The silver-blue shimmering fish were simply too fast, and the technology was not designed for them. "We only got the knack for it in 2007," Árnason explains.

The fisherman realized that they needed boats with more powerful engines, to allow them to haul hefty nets and move fast enough to catch the mackerel—much faster than they were used to sailing when in pursuit of cod, haddock, and caplin. They also had to develop new product ranges and give marketing a boost. They succeeded: By 2008, Iceland's fisherman had caught over 110,000 tons of mackerel.

The Faroe Islands also had their fill of fish, which were now gathering along their coastline. As far as they were concerned, they had exclusive fishing rights in their own waters, much to the annoyance of Great Britain—the mackerel were the Brits' most important fish stocks and ensured the livelihoods of many fishing villages in Scotland. They accused the Icelanders of stealing "their" fish.[14] They had had a score to settle with Iceland since the so-called Cod Wars. In December 1975, after cod stocks collapsed, Iceland unilaterally extended its fishing zone from 50 to 200 nautical miles. Both Britain and West Germany refused to withdraw their fleets from their traditional fishing grounds, so Icelandic coastal patrols fired warning shots at the foreign boats and employed "clippers," their own invention, to cut forty-six British and nine German trawler nets. However, when Iceland—a country of geostrategic importance—threatened to pull out of NATO, the Brits were forced to back down.[15]

Now, Norway and the EU would have to catch less in order to maintain overall mackerel stocks—something they refused to accept. As a result, fish stocks were overfished with no consideration for the recommendations of the International Council for the Exploration of the Sea (ICES). In 2012, the inevitable happened: The MSC label was withdrawn from stocks in the North Atlantic, a bitter blow for marketing.

Brussels and Oslo decided to put an end to it. They prepared to impose sanctions.

A BLACK BOX IN A MACKEREL'S EAR
Bremerhaven, January 23, 2020

It's like something out of a horror film. Aprons hang from a hook on the wall. In the middle of the room, there is a brass table, covered in pools of blood. On it lie two bloodstained knives—a kitchen knife and a longer blade. Mackerel are piled high on the table. They have come from the very north of the North Sea.

Two young men have donned aprons and are defrosting the fish in the cellar of the Thünen Institute for Sea Fisheries in Bremerhaven, lining them up, weighing and measuring them. They then slit the fishes' bellies lengthwise, remove the gonads, and check them. They cut down diagonally into the head and fold it out using tweezers to poke around in a brain sac filled with liquid; they're looking for the ear stone, the otolith. This stone governs the fish's sense of balance. Scientists can use it as one would an airplane's black box: It enables them to tell, for instance, how much the fish has eaten. The less it has eaten, the less calcium the stone will have absorbed, and the more translucent the stone will be. But the stone also enables scientists to re-create the fish's migration routes. "In essence, it reveals everything that the fish has done," says Gerd Kraus, director of the institute.

A mackerel's otolith is so small as to be hardly perceptible, whereas a cod's is as big as a hazelnut. Kraus reaches into a bowl and pulls out one of the little white stones. He's especially interested in the age of the fish, evidence of which can be found in the stone, more precisely, in a triangle that reveals the years of the fish's life in cross-section, like the rings of a tree. "We're recording the demographics of fish stocks," says Kraus. "This one lived to three years of age."

Age is crucial. It can be used to calculate mortality rates and establish a population pyramid. Using computer modeling, it's even possible to tell what whole age groups died of—be it fishing or natural causes. Scientists use this data to predict how fish will develop in years to come and make recommendations as to how many fish fishermen should be allowed to catch in Europe's seas. Whether these recommendations are implemented is another matter altogether—as the story of the mackerel shows.

NO ENTRY FOR SHIPS FROM ICELAND

In August 2013, the EU disallowed imports of mackerel from the Faroe Islands. What was more, fishing vessels from this autonomous group of islands were no longer allowed to dock in European ports. The two sides were only able to reach an agreement the following year—the Faroe Islands would now be able to catch significantly more fish than had been agreed to in the old regulations.[16]

The EU and Norway also imposed sanctions on Iceland. Norway had banned fishing boats from Iceland from entering its ports back in 2010, as had some sites in Scotland. Scottish fishermen lined up their tugboats to block the trawlers from Iceland and the Faroe Islands.[17] In turn, Iceland prohibited Scottish haulers from fishing for cod and haddock in its waters, impacting fishermen who had had nothing at all to do with the tussle over mackerel.

The row imploded; there seemed to be no end to the haggling over catch percentages. It became increasingly clear that the international regime regulating fishing rights in the oceans was not fit for purpose. Generally speaking, if fish stocks cross national borders, they find themselves in a legal vacuum. The United Nations Convention on the Law of the Sea encourages nations to cooperate and enter into new agreements, but years can go by before these are successfully negotiated. In the meantime, fish stocks are being massively overfished.[18] US marine biologist Malin Pinsky compares the situation to two children gobbling up a cake the moment it is placed in front of them, leaving hardly a crumb behind. "We're not well prepared for species on the move in the ocean," says Pinsky, a professor at the Institute for Ecology, Evolution and Natural Resources at Rutgers University in New Brunswick, New Jersey. "Our fisheries management is largely based around the idea that species more or less stay where they've been historically."

Perhaps it has something to do with the fact that almost everything that plays out beneath the water's surface is hidden from view. We can't see the great migrations that are underway in our oceans right now. Take the North Sea, for instance: If species such as cod and mackerel are migrating north, it doesn't mean that the North Sea is emptying; new species are moving in from the South. Scientists speak of the "subtropicalization" of the North Sea. Where cold-water species like herring, sprats, and cod were once dominant, warmth-loving species are increasingly taking the helm. Red mullet and anchovy are two such examples. In the mid-eighties, these two species were still practically unheard of in the North Sea; today, they inhabit a significant portion of it.[19]

Fishermen in Germany are similarly puzzled, increasingly finding tuna fish and sardines in their nets. Meanwhile, these species are finding their way onto shop counters, as the practice of throwing bycatch overboard was banned a couple of years ago. Some British fisheries have gone so far as to pivot their operations to focus entirely on sea bream and red mullet.[20] However, these newcomers have yet to make up for the loss of larger and often more profitable cold-water species.

These changes are particularly noticeable in squid. In the early '50s, this species was never found in fishermen's nets, whereas now, some years will see fishermen catching over 3,300 tons of squid across the North Sea. Squid and octopus are among those species that are benefiting from climate change. As the North Sea warms, they are able to grow faster, enabling them to sooner reach a size at which they are safe from predatory fish such as cod. Favorable conditions have helped the squid to turn the tables on their old foes—and hunt young cod themselves.

"SALTED FISH IS LIFE"

The UN Convention on the Law of the Sea governs how nations are to behave in the event of fish stocks crossing national borders. They are obliged to consider a range of questions: Which countries have fished a species of fish in the past, which countries depend on them economically, and where are the fish now?

In the row over the mackerel off the coast of Iceland, however, it became clear just how small a part these considerations had to play in the settlement. The individual parties did take the laws very seriously, particularly those that served their own interests. Norway, for instance, held the view that the mackerel stocks should be divided according to the quantities present in each "exclusive economic zone"; after all, there were all kinds of mackerel splashing about off their own coastline. By contrast, the EU invoked the principle of historic fishing rights: A nation that had always caught a certain species ought to be able to do so in the future—regardless of that species' migration. Iceland came out strongly in favor of the principle of economic dependence on fishing, as its economy was undoubtedly dependent upon it. "Salted fish is life," claimed the Icelandic writer and Nobel Laureate Halldór Laxness almost one hundred years ago. In 1944, the fishing sector led Iceland to independence from Denmark, and, after the Second World War, it paved the way for modernity and brought prosperity.[21] Today, almost a fifth of Iceland's economy is tied to fishing.[22] While other sectors, such as tourism, have seen a sharp increase in overall economic significance, fishing is often the only industry in areas outside of the capital city. "Many villages and towns along the coast depend exclusively on fishing," Árnason explains. "They simply don't have anything else."

Due to the woolly definitions set out in the UN Convention, nations have been able to interpret the laws, assigning importance to them however they wished. An impasse was approaching. And another factor made conciliation difficult: Marine scientists were unsure how large mackerel stocks were, what had led them to spread northward, and whether they would remain there in the future. For a time, the EU and Norway completely refused to accept that mackerel had encroached on Icelandic waters.

Postdoctoral researcher Jessica Spijkers from the Stockholm Resilience Centre wanted to find the underlying cause of the conflict and interviewed twenty-six politicians, businesspeople, and civil society representatives from Norway, the Faroe Islands, Iceland, and the EU who had been involved in the dispute.[23] One Icelandic trader related the following absurd dialogue, told as it played out over the bargaining table in 2008:

"There's no mackerel in Icelandic waters!"

"There must be mackerel there, we have been catching mackerel!"

"Well, it's probably more herring that you claim is mackerel."

Soon, however, there was no denying that mackerel were bustling about in the waters off Iceland. Norway changed tack and claimed that the fish had migrated northwestward only temporarily and that they would eventually return to their original habitat. On such grounds, they claimed, Iceland should not be granted any permanent fishing rights. "Accepting that the shift is caused by climate change would confirm the permanence of the shift," Spijkers explains.

A turning point was reached in 2012. Brussels was ready to back down and grant Icelanders a sizable proportion of the mackerel catch. There were two reasons for this. The first was that scientists now agreed that Icelandic waters did, in fact, contain large mackerel stocks. In the intervening period, mackerel had also appeared off the south coast of Greenland and even off Svalbard in 2013.[24]

The second reason was that Iceland had applied to join the European Union. Brussels wanted the matter off the table before it entered into official negotiations with Iceland. And so, the federation offered Iceland a catch quota of 11.9 percent of mackerel stocks. Iceland accepted. They would seal the deal in Edinburgh in March 2014.

It all seemed to be going swimmingly.

But they had forgotten to ask the Norwegians.

HIGH-CONFLICT POTENTIAL

The mackerel were just the beginning. Just after the turn of the millennium, shoals of anchovies appeared in the English Channel and the southern North Sea. The spindle-shaped fish with their forked tailfins landed in the nets of British trawlers, much to the annoyance of the French and the Spanish, who had long since overfished stocks in the Bay of Biscay—and now wanted exclusive access to the new fishing grounds. They were their stocks to begin with, the argument went.

Genetic analysis revealed that the anchovies off the south coast of England were probably left over from previous fish stocks in the west of the Channel and were now benefiting from climate change and spreading up into the North Sea.[25]

If the arrival of individual species can lead to conflicts of this kind—and, notably, between democratic neighbors with decades of friendly relations behind them—what kind of political conflict could emerge if fish stocks across the entire planet shift in line with global warming? Consider, for example, regions such as Southeast Asia where conflicts over marine law are still smoldering. Disagreements over migrating fish stocks could come to a head; the fishing industry has already sustained losses in productivity as a result of climate change.[26]

Malin Pinsky has calculated how the habitats of around nine hundred commercially significant species of fish and marine invertebrates will shift by the end of the century. Many nations could encounter between one and five new populations in their exclusive economic zones if the world continues to emit greenhouse gases at current rates. For some countries in Asia with especially high conflict potential, this could rise to as many as ten new populations. This is not an exact prediction, Pinsky admits, but this analysis makes clear what is in store for us if the world does not come together and cooperate over its marine ecosystems.

CONSEQUENCES FOR ICELAND

Not having been consulted, the Norwegians felt that the agreement between the EU and Iceland had been made behind their backs, and so they let the deal fall through. They also ensured that Iceland was banned from European fishing grounds and forced the Faroe Islands—Iceland's longtime ally in the mackerel dispute—to agree to prevent Icelandic fishermen from catching mackerel in its waters. "Icelanders feel they've been abandoned," says Árnason. "They feel they are in the right and they have to recognize that Norway and the EU are using their power to force us to give up any hopes of a catch."

The Icelanders' protest against Norway fizzled out. The bigger, more powerful neighbor had won—for the time being.

It was the mackerel themselves that prompted a temporary détente in the crisis. Their stocks had grown, experts from ICES explained; all countries would now be permitted a larger catch.

The solution was anything but sustainable, and the political damage had been done. The dispute led to Iceland's EU accession process being shelved. Eventually, Icelanders held a referendum in which they went so far as to vote against membership of the EU—not least because of the controversy over fishing.[27]

What would a prudent fishing policy look like in the age of global warming? Pinsky and two dozen of his fellow scientists posed this question as part of a research group. First, they concluded, all nations would have to use their fishing grounds sustainably, adhering to scientific recommendations. "Because if there aren't any fish left, then there's nothing left to share," says Pinsky.

Second, the scientists agreed, nations would need to agree on how to divide up fish stocks if they cross political borders. International courts could be called upon to navigate this issue; nations could receive compensation if their fish stocks were lost, and flexible trade with catch rights, as practiced by some Pacific Island nations, could also help. According to this idea, if a country has greater access to fish stocks, it can borrow quotas from other countries. "Systems like that could help fisheries respond more rapidly to changes in species distribution," says Pinsky.

Ideally, nations would reach an agreement before species shift their ranges of distribution altogether. "It's so much easier to address these sharing mechanisms when they're not in the heat of a conflict," says Pinsky.

"WE'RE WINNERS AND WE'RE LOSERS, TOO"

In 2019, the mackerel wars threatened to flare up again. In March, mackerel fishermen in the Northeast Atlantic once again lost the MSC seal for sustainable fishing, as stocks had almost halved since their high point in 2014, falling below a sustainable minimum population size.[28] Despite this, Iceland announced that it would once again be raising its quota of its own accord—from 108,000 metric tons to 140,000. Russia and, more recently, Greenland also caught mackerel in their waters without permission, such that the other members of the Northeast Atlantic Fishing Commission were obliged to avoid catching greater quantities of fish in order to stay within the scientific guidelines.[29] The EU and Norway now wanted to put a stop to Iceland's activities, threatening the country with sanctions again in the summer.[30]

Yet the island nation could not be swayed. There were very practical reasons for this. Iceland was hit especially hard by the financial crisis that gripped the globe in 2008 and was pushed to the verge of bankruptcy, leading to a resurgence in traditional industries like fishing. Icelanders had put together a large fleet of trawlers at great expense. Yet the forty-million-euro ships suddenly lost track of important fish species: blue whiting, for instance, which had been hopelessly overfished and migrated northwest to the waters off Greenland. Caplin was another: Great shoals inhabited the waters surrounding the island and had done so for as long as Icelanders could remember. Until recently, caplin, which tastes similar to herring, was the second-most significant species in terms of exports; fishermen pulled more than 1.7 million tons out of the ocean each year. In 2018, however, the caplin seemed to disappear from the coast of Iceland without a trace. Local fishermen claim to have observed them migrating north to colder waters, similar to what Canadian scientists had seen in West Greenland.[31] Icelanders now feared for the fish they relied upon most— cod—as these primarily feed on whiting.

Some also blame the new arrivals: mackerel. Mackerel are voracious predators, eating anything smaller than themselves, such as sand eels, on which sea birds, haddock, cod, and whiting also feed. But the mackerel are not just stealing a source of food from these commercial fish species but also hunting their young. "They were causing absolute chaos in the marine ecosystem," says Árnason.

He witnessed it for himself when taking his grandson fishing off the coast of Reykjavík. "It was impressive to see how they consumed everything right up to the coastline," Árnason recalls.

Once they've fattened themselves up, the mackerel journey up toward the Faroe Islands and the coast of Norway. "When it comes to climate change, we're winners and we're losers, too," the director of a fishing company in Vestmannaeyjar in the south of Iceland tells me. Winners, because profitable species of fish such as mackerel and cod are moving north. Losers, because these predatory fish help themselves to local stocks of other fish species and react with great sensitivity to the warming of the seas.[32] "The world is changing and we've got to keep up."

Scientists at the Alfred Wegener Institute for Polar and Marine Research in Bremerhaven have been investigating what will happen if the oceans on Europe's doorstep continue to warm and acidify. They caught cod in the southern Barents Sea and brought them into the laboratory. In the lab, they exposed the fish to various temperatures and high concentrations of CO_2, according to different climate scenarios for the year 2100. Their findings showed that if the Earth warms by more than 1.5°C (2.7°F), it will exceed the critical threshold for current spawning grounds. If greenhouse gas emissions are not brought under control, by the end of the century it will be impossible for cod to spawn south of the Arctic Circle. The major fishing grounds of today, such as those off the coast of Iceland and Norway, would be lost forever.[33]

The future is uncertain, and this is something that Icelanders understand. A species of fish can disappear from one day to the next—or even reappear. As for Árnason and his grandson, they made the best of it and caught a few mackerel off the coast of Reykjavík. "They're very fatty fish," he says. "Perfect for barbecues."

6

—

It's Heating Up

Before we continue on our journey from the North Pole into the tropics, let's pause for a moment and consider who exactly the Earth's plants and animals are taking their lead from when they are moved to migrate. Is there a pattern, once they shift toward the poles and up into the mountains?

A spot in Berlin provides a natural starting point for diving into this question, and I'm keen to visit. I jump on my bike and ride over to Jägerstraße, just a few blocks from my office on Friedrichstraße. I'm looking for the birthplace of one of the world's most renowned and beloved Germans. Over two hundred years ago, he was the first to discover how life on Earth was distributed—in other words, why species of animals and plants came to live where they do.

At Gendarmenmarkt, I turn off into Jägerstraße. Number 22 is a massive building, covered in scaffolding. I dismount and wheel my bike along the sandstone wall, sheltered by steel boards that keep off the summer sun, until I reach a plaque between two windows, its brass surface streaked with green. A striking face stands out; it has a high forehead, a steadfast gaze, and a heavy bottom lip. The text underneath reads:

ON THIS SITE STOOD THE BIRTHPLACE OF THE GREAT GERMAN
NATURAL SCIENTIST AND MEMBER OF THE ACADEMY OF
SCIENCES, FRIEDRICH WILHELM ALEXANDER VON HUMBOLDT

A bang echoes from above, making me wince. One of the builders must have dropped something heavy; there's a clatter and a crash. Cars and scooters roar past. This part of town would have been much more peaceful when Humboldt entered the world on September 14, 1769. At the time, the Baroque Domestikenhaus next door was home to clerks and attendants in the court of Friedrich Wilhelm I, until the Maritime Trading Company moved in in 1777, eventually purchasing Humboldt's childhood home. A few streets north in what was once the suburb of Oranienburg, clouds of smoke swirled up into the sky from the August Borsig engineering plant. At the time, Berliners called this industrial nucleus "Feuerland," or "the Land of Fire."[1] Yet Humboldt was less interested in the industry around the corner than he was in the nature to be found farther out. In 1799, he embarked upon an expedition to South America that would revolutionize our understanding of the natural world.

On July 23, 1802, Humboldt, together with the French botanist Aimé Bonpland, climbed Chimborazo, a volcano in Ecuador standing at over 20,000 feet, which was considered to be the world's highest mountain at the time. "The trip from Quito to Chimborazo was like a botanical journey from the equator to the poles, but vertical: a whole world of plant life, piled layer upon layer," writes historian Andrea Wulf, in her fascinating biography of Humboldt.[2] "One zone of vegetation after another, the higher they climbed, from the tropical species in the valleys to the last scraps of lichen just beneath the snow line."

It must have been quite a difficult climb for Humboldt, a Prussian aristocrat. "Dazed, half frozen and gasping for breath in the thin air, Humboldt and his small entourage crept on their hands and knees over steep ridges and razor-sharp stones," Wulf writes. But this torture was not for nothing. Humboldt now understood what held nature together. A "glimpse into the living organism of the Earth" was how he would describe it a few years later.[3]

Humboldt had meticulously categorized all collected species according to their respective locations above sea level and charted them using barometers, hygrometers, and electrometers, even capturing the blueness of the sky with a cyanometer. Out of this emerged the famous profile of the Andes, which Humboldt created in February 1803 in the heat of the Ecuadoran port of Guayaquil. It was from here that he looked out at the gigantic expired volcano while, behind it, the still-active Cotopaxi constantly sent its "crashing underground thunder" his way, resounding in his ears like "the thundering of heavy artillery." In this profile, he assigned an altitude level to all the plants he had found—from fan-leaved palms at the foot of the volcano to deciduous trees, shrubs, and grasses to the mosses and lichens beneath the snowcapped summit. He then furnished them with physical data. This itself was a new way of describing nature, but at the time botanists were concentrating more on finding, describing, and classifying new species and less with connecting them with their surroundings, and they certainly weren't concerned about creating a new system out of this. Thus, Humboldt saw himself as creating a "geography of plants," "a discipline for which a name barely exists." "It observes plants according to the proportions of their distribution across different climates."

On the climb, Humboldt repeatedly encountered plants similar to those he had seen in the Alps and Pyrenees. He observed them closely and compared the environmental conditions in which they grew. He concluded that there must be a great force that united them across continents: The world's vegetation zones corresponded to its climate zones. Today, such knowledge seems obvious, but it was revolutionary at the time. It was this that saw Humboldt recognized as the founder of biogeography.[4]

In 1817, Humboldt distilled his ideas into a map similar to the ones we would recognize today in an evening weather report. It illustrated what was otherwise impossible to see:

bands spanning the world, connecting places with the same mean temperature. Humboldt called these *isotherms*. The term originated from Greek and combined the words for "same" and "heat." On his map Carte des Lignes Isothermes, Humboldt sketched the bands of heat that connected certain regions of America, Europe, and Asia. According to the map, mean temperatures are similar in north Florida and Naples, Italy, as are temperatures in Boston and Stockholm. The isotherms drift around the Earth, sometimes thicker, sometimes thinner. Despite their being invisible, the world's plant life naturally orients itself along isotherm curves. Humboldt considered this system to be reasonably fixed; after all, "each altitude is assigned its own, unchanging temperature."[5]

But they are not truly unchanging, as we now know. Even a visionary like Humboldt couldn't have imagined how these temperature bands have shifted up toward the two poles, up mountains, and how various species are hot on their heels. Even

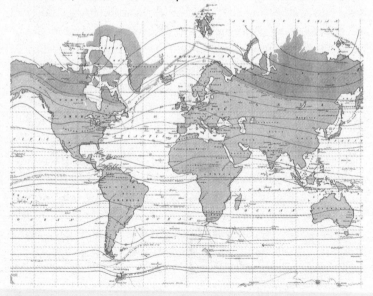

First illustrated in 1817 by Alexander von Humboldt, isotherms are the key into the world of biogeography, a scientific field that identifies patterns of plant and animal distribution and, lately, has shown how climate change has forced a shift toward the poles.

on Chimborazo, two hundred years later, botanists analyzed vegetation and demonstrated it had migrated an average of over 1,600 feet upward since Humboldt's day.[6] Humboldt would of course have considered periodical fluctuations between periods of cold and warm in Earth's history, as well as the migration of species, after elephant tusks and the skeletons of tapirs and crocodiles were uncovered in Europe.[7] Humboldt was also the first person to recognize that our species is capable of changing the climate: "by felling forests, altering the distribution of bodies of water, and through the development of great masses of vapor and gas around centers of industry."[8] But he was not, at that time, able to envision the extent to which humans would exploit coal, oil, and natural gas, changing the proportion of molecules in the atmosphere and, in doing so, shifting the positions of the isotherms.[9]

Someone who did envision this, however, is Jonathan Lenoir. Together with his colleagues, Lenoir, a biostatistician at the University of Picardie Jules Verne in Amiens, northern France, spent five years examining studies, and learning about each recorded species. Lenoir painstakingly digitized the team's findings in a database that he calls "BioShifts." Never before has anyone mapped shifts in species so comprehensively[10] and in such detail. Lenoir was then able to compare this information to the shifts in isotherms to see the big picture.

The thermal belts are shifting toward the poles by just over half a mile per year on average[11]; in doing so, they determine the speed at which species have to react. According to analysis, it is the sea-dwelling species that have managed this most successfully thus far. "We were expecting that they were moving faster than land species," says Lenoir. "But not with such a big difference at the end."

On average, plankton, fish, and whales are retreating toward the poles at a rate of about 4 miles each year, while land-dwelling species are migrating only 1.2 miles each year—on average—which means that many species are moving considerably slower and falling behind their usual zones of temperature. This places them under threat: They may die out before they are able to spread into cooler climes. There was something else Lenoir spotted when analyzing his data: Some species were reacting to climate change completely differently. Some did not migrate at all and some even migrated in the wrong direction, toward the tropics or down mountains. Lenoir wanted to find out why this could be. For a long time, he couldn't make out any particular pattern. Only when he included humans in the modeling did the image become clearer: The more the oceans heat up *and* the more pressure human beings exert on ecosystems and overfish the oceans, the faster species flee. "These factors create a synergy, accelerating the rate at which species migrate," says Lenoir. "Marine species follow their isotherms very closely."

On land, the situation was turned on its head. In places where humans interfere particularly heavily in the landscape, covering it with roads, settlements, and fields for agriculture, they restrict the possibilities for animals and plants to move. Unlike sea-dwelling creatures living in the three-dimensional ocean, these species often have very few options when it comes to escaping; nature places all kinds of obstacles in their way, from mountains and rivers to continental margins.

Consequently, over the course of Earth's history, land-dwelling species have, to a greater extent, had to learn how to deal with fluctuations in climate, adapting to new conditions without immediately trying to flee. They can do this by naturally selecting themselves according to their genes and outer shape,[12] or by seeking refuge. Forests, for example, provide countless species with shade, creating a microclimate beneath their canopies that can be as much as 11°F cooler than a nearby meadow. "Species use this the way we use our homes to shield us from heat or cold," says Lenoir. "And this may contribute to us observing many fewer migrations on land."[13]

Yet the longest-living organisms in the world—symbols of permanence and stability, keystone species whose leafy canopies offer countless species refuge and shade,[14] protecting them from climate change—stand in stark contrast to speedy mackerel. Time to look beneath the canopy and explore the world of trees.

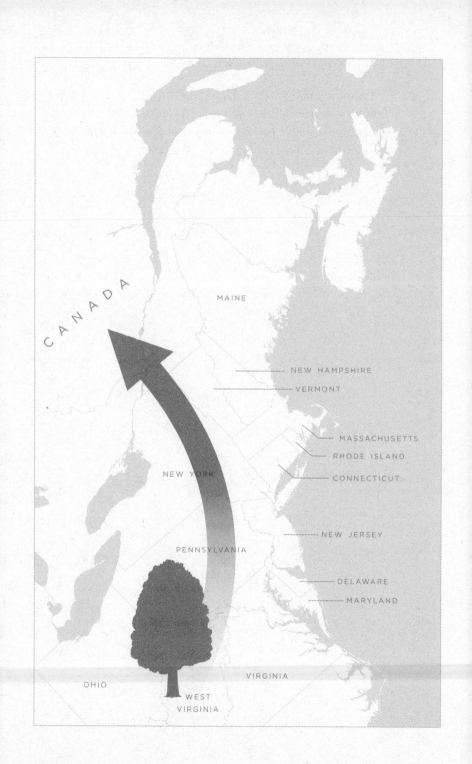

The Forests Are on the Move

The first attempts were not very encouraging, almost pathetic—he knew that himself. The syrup tasted a little burnt, even inedible; maybe he'd overcooked the birch sap, or maybe he'd used the wrong equipment to harvest it. Like most New Englanders, David Moore, a tree physiologist in Lee, New Hampshire, grew up on delectably flavorful maple syrup; the viscous mass he'd wrought didn't even come close. But he was on a mission and not easily dissuaded from his plan: He was convinced he could avoid the standard sugar maple and still make delicious syrup out of any other local tree species—it was just a matter of uncovering their quirks and nuances.

When he walked through the woods, he tapped hickories, apple trees, sycamores, birches, and beeches. Each time, the man with the short, black hair, round face, and dark eyes drilled a little hole into the trunk, inserted a tube, and let the sap flow into a metal bucket—like milking a cow.

Then he poured the sap into plastic tanks, loaded the tanks onto the bed of his truck, and drove back to Lee, where he boiled the sap in steel pots over a wood stove in his driveway. Later, he would build his own sugar shack on a little farm, where he filtered the sap, steamed it off in the evaporator, and concentrated it.

In 2008, while Moore was studying plant sciences at the University of New Hampshire, he started a project that aimed to measure the sugar content of different species of birch, a prolific producer of sap but with lower sugar content and a shorter harvest season than for sugar maple trees. In the hierarchy of syrup-producing trees, birches, according to Moore, rank second, right behind the undisputed champion sugar maples and above the promising walnuts, beech trees, and sycamores. After determining the birch species with the highest sugar content—paper birch, which has the added advantage of producing sap for a couple of weeks after sugar maples are finished for the season—he felt ready to start a business, beginning with birch but also branching out to sycamore, walnut, and beech syrup. He called his company the Crooked Chimney. It took him two years of hard work until, eventually, the dark and thick masses he produced flattered his taste buds at last.

Still, Moore knew it was an almost impossible task to compete against maple syrup. *Acer saccharum* simply has so much going for it. In winter, it yields a good deal of sap from which the syrup is made. This high output, paired with its high sugar content and long harvest season, makes it no surprise that maple syrup has dominated the market since Europeans settled in New England and had carried on with a syrup-making process the Native Americans had invented.

And sugar maples' advantages extend beyond syrup. They dominate many northern hardwood forests and have become part of the identity of the region, bathing the autumnal New England landscape in yellow, orange, and red as their leaves change color—a major tourist draw. Sugar maples are also an important source of saw timber, particularly for construction and furniture manufacturing.[1]

Put that all together, and sugar maples are a vital economic powerhouse for states whose primary natural resources are their forests. In 2021, the US produced almost 3.5 million gallons of maple syrup, with Vermont leading the way with 1.5 million gallons—worth roughly $130 million altogether. When spring awakens, you can see rising steam coming out of the chimneys of small cabins spread all over the woods in New England. Here, families can practice the art of tapping sugar maple trees and producing syrup—and then, of course, they'll taste the new vintage. "Visitors come from all over and outside of New England to have that cultural experience of trying the syrup once it's been produced and participating in the whole process," says Heidi Asbjornsen, a forestry professor at the University of New Hampshire.

"It's got this mystique about it," says David Moore.

Which raises questions about Moore's quest: Why look for an alternative when sugar maples are seemingly already the perfect solution?

It's not only about new markets or expanding the production season beyond the six- to eight-week window in February and March when sugar maples in New Hampshire make sap. It's because the sugar maples' fate has become so tenuous.

Climate change has already shifted the sap-harvest season in some years to as early as December—a significant planning challenge for producers. But there is an even bigger threat in the long run: The more Earth warms up, the farther north the region of maximum sap flow. By 2100, this area is expected to shift 250 miles northward, according to scientists from the University of Massachusetts Amherst and other institutions in the US and Canada.[2] They project that while this means higher output in Canada, sugar maple syrup production in most of New England will drop by half. "Projections indicated dramatic and mostly negative changes in syrup production per tap by the end of the century," the study says. Virginia and Indiana are expected to lose virtually all syrup production; scientists expect output in Massachusetts and New Hampshire to be cut by half.

The reason for this shift can be found in the idiosyncrasies of the freeze–thaw cycle. Sugar maples first need bitter cold temperatures to "stimulate both sugar formation within the stem and sap flow during spring time," as a recent paper explains.[3] Then, to collect the sap, nights below freezing must alternate with days between 37 and 45°F "to create a positive pressure to cause sap to flow down from tree branches and out the tree through a tap and collection system." Such repeated freeze–thaw periods are becoming more and more unlikely in the southern part of the range of sugar maples, as the winters become warmer and spring arrives earlier.

The question is: Have sugar maples already been set in motion because of climate change?

NOTABLE SHIFTS NORTHWARD

The first to find evidence of a northward shift was Christopher Woodall. An avid hiker since he was a boy, Woodall loved trekking through forests that seemed to him as if they had existed forever. He was ten years old when he had an eye-opening experience while walking through Great Smoky Mountains National Park. The park was founded in 1934, so the young Woodall expected to see only old-growth forest. But, to his surprise, he also noticed pockets of young tree species roughly his own height; they must have established themselves after a disturbance such as a storm or acid rain had wiped out a section of mature trees. Suddenly, he understood that forests are more dynamic than he'd thought. "They shift like a mosaic through the landscape," Woodall told me.

But it was not until the early years of this millennium that Woodall became aware of tree species in the northern US that were not only migrating but doing so in a particular pattern— one shaped by climate change. He'd just finished his silviculture studies at the University of Montana and had begun working for the US Forestry Inventory Program in Saint Paul, Minnesota, which had just gone through a revolution in the way it tracked forests: For centuries, forests in the US hadn't been traced a systematic way—all that existed were range maps from the 1970s created based on personal observations.

Then, in 2000, everything changed. Now, every year a whole legion of employees in khaki-colored clothes swarmed through all kinds of forests and measured them in 130,000 distinct plots across the country. This opened up a brand-new world of research possibilities for ecologists. "I was amazed of so much data," Woodall recalls. But soon, he realized that his colleagues hardly looked at it. "We typically only used it for doing timber resource assessments," he says, and there is still a bit of lingering disappointment in his voice.

He knew that climate change had already set the natural world in motion, as initial studies had described this phenomenon in different parts of the globe. He decided to try to find out whether this included the forests of the eastern US: not by putting on his hiking boots and trekking through the woods, but by walking through virtual forests—that is, by looking at the data. "Most of the ecologists were studying a couple of individual research plots," he says. "I'm different. I'm a data person."

In other words: He wanted to see the big picture.

Every day, after his regular work, he ran algorithms through the national database on an old desktop computer in his federal lab in Saint Paul, without his supervisor knowing. The problem was that the time span of the new data set was too short to uncover any range shifts of tree species. So, he tried a different source: tree demography, an inventory that listed not only where mature trees were standing but also where new tree seedlings were growing. By comparing both for individual species, Woodall hoped to see a pattern. And he did. "I was shocked to find how different the tree regeneration is, spatially, for some species compared to where they currently are," he recalls. "Their regeneration was shifting north, potentially ahead of the adult trees."

He wrote a study with colleagues from different research stations of the US Department of Agriculture and published it in a scientific journal in 2009.[4] It concluded that "the process of northward tree migration in the eastern United States is currently underway with rates approaching 100 km (62 miles)/century for many species."

This phenomenon was even more pronounced for northern tree species, where the landscape was warming at the highest rates. There, the mean latitude of seedlings turned out to be, on average, more than 12 miles farther north than the mean latitude of tree biomass. One of the species that already showed notable shifts northward was the sugar maple in New England. "Regeneration is more abundant and prevalent in the northern part of its range compared to its southern part of its range, where it is very sparse," Woodall explains.

That doesn't mean that the whole maple syrup industry and all the sugar shacks in New England are going to disappear overnight. Sugar maple trees grow much older than humans—some more than four hundred years. At the end of the century, many of them probably will still be digging their roots into the soils of the Northeast. Additionally, there are more than half a billion sugar maple trees that are too young today for tapping but will be mature enough sometime this century. At least in the short term, sap harvest could even increase, partly because of technologies such as new vacuum tubing systems.[5]

But nobody knows how increasingly severe droughts and a shrinking sap-flow season—at least at the southern edge of the sugar maple's distribution range—will play out. Droughts in New England have been almost nonexistent for decades, but this is changing: In 2016 and 2020, the region experienced abnormally dry conditions, and many tree species suffered tremendously. In the long run, many states in New England could lose their beloved sugar maple sap.

Lee, 2009

In the first year of his quest for new syrup sources, David Moore tapped two dozen trees. The next year, he tapped about a hundred. Ultimately, he tapped nearly 250 trees. In the course of his experimentation, he discovered the best time to tap, the appropriate temperatures for boiling, and the ideal equipment material (stainless steel).

When he'd finally developed his birch syrup's uniquely savory flavor, his customers found it stronger and more bitter than sugar maple syrup—almost smoky—and detected notes of raspberry and molasses. Moore describes the taste of his sycamore syrup as something like butterscotch or honey. He ran the Crooked Chimney for six years, selling his syrup to restaurants, farmers markets, and over the counter at 11 Randall Road in Lee. Every year, shortly after the harvest season in March and April, he sold out of the birch syrup.

It's clear, then, that if the sugar maples were to vanish from New Hampshire, there are successors in place. But it would be a dramatic cultural shift for New England—and not the only one of its kind that looms ahead in the US.

Other tree icons in America are migrating, too, and with them, many ecosystem services. Take the oak-dominated forests of the central US: The ranges of species such as white oak and bur oak are shifting and shrinking, mainly because of human intervention and indirect effects of climate change, including increasing threats from insects. The timber of these white oaks is not only important for furniture and flooring, but also for barrels for the wine and bourbon industries.[6] "The character of that industry will change," Christopher Woodall predicts.

In other regions of the temperate zone, important tree species are migrating, too, and thus pose major challenges for the affected countries. In Germany, the most important species for timber production—the Scots pine (*Pinus sylvestris*) and Norway spruce (*Picea abies*)—suffered substantially after three very hot and dry summers in a row beginning in 2018. Particularly hard hit were the spruce trees, the needles of which discolored from the lack of water before the bark beetle attacked the weakened trunks. In the Harz Mountains, Saxon Switzerland, and the southern Black Forest, forest skeletons gaped.

The Scots pine hasn't fared much better. In just a few years, entire forests have disappeared along the southern Rhine plain as far as Frankfurt. All that remains is grass steppe. According to computer models,[7] the Scots pine is likely to lose more than half of its range by the end of the century and retreat from central and southeastern Europe to the higher elevations of the Alps and the Carpathians as well as northern Europe—similar to the Norway spruce, which is predicted to largely withdraw from western, central, and eastern Europe and migrate to northern Europe.[8] Marc Hanewinkel, a professor of forestry economics and forest planning at the University of Freiburg, estimates that the economic costs in Europe would amount to several hundred billion dollars if profitable coniferous woods disappear and Mediterranean oak species replace them by the end of the century.[9]

Let's have a brief look at the predictive power of these computer models. They enable a glimpse into the future: In order to find out where climate change might shift the habitats of a species, scientists first have to determine its "climate envelope." Most of the time, they just look at where the species is currently and what the climatic conditions are like—how warm and humid it is. In the second step, they calculate where the climate niche will migrate in the future by running through various climate scenarios.

These correlative species distribution models, as they are known, are not undisputed. For one, a species may not be able to make full use of the habitat that the climate makes available to it. Perhaps portions of it are occupied by human settlements or agricultural land—perhaps a mountain or a river—or a more competitive species contains their spread. The more these additional factors can be accounted for, the greater the accuracy of the model.

The species in question may also not find itself in an equilibrium with its current climate. This means its "realized niche" is not its "fundamental niche"—a species' absolute climatic limits. As a result, the future areas of distribution posited by scientists could turn out to be somewhere quite different.

Nevertheless, most biologists prefer this method of computer modeling to others.[10] It has the invaluable advantage of being applicable to a large number of species with relative ease and efficiency. And it serves as a practicable starting point—a first risk assessment that cannot, however, be confused with the migrations that actually emerge.[11] Sometimes, the shifts in reality are much more dramatic than the predicted ones, surprising even experts like Woodall.

After persistent extreme hot and dry years since the 1990s, whole forests in western North America have vanished, leaving ground for grasslands and shrublands. Fir, spruce, pine, ash, and poplar died either directly because of water stress or by the attack of pests and fire—or by a combination of these factors. In many cases, the dieback was most pronounced at the edge of their distribution, where climatic conditions prevailed under which they had just been able to survive—up to now. "Due to drought and heat stress, we are seeing higher mortality rates in all forest types globally that are probably unprecedented in their history," the forest ecologist Craig D. Allen, who introduced the term "hotter droughts," told me. "The atmosphere is literally pulling the water out of the vegetation."

At the same time, tree species were sprouting in the far north, where they couldn't exist before because the climate conditions had been too harsh.

FORESTS ON THE MOVE

A few years ago, Woodall, now working for the Forest Service in Durham, New Hampshire, collaborated with forest specialists from different universities in the US to find out if the phenomenon of tree migration would also apply to whole forests—not only to particular tree species. In the time since his initial research, the data sets had grown tremendously, so

Woodall and his colleagues had to experiment with new methods, such as machine learning, to master them.

The results from Woodall's investigation haven't been published yet. But what he and his fellow scientists have discovered so far has surprised them: In some cases, entire forests are changing type—a woodland of oaks, say, becoming a forest dominated by maples.

It's a dramatic shift, but it also reveals how even minor shifts in individual tree species can cause a ripple effect. "In many cases all it takes is just another two or three percent of one species to tip a forest over into another type of tree community," Woodall explains.

Imagine a huge hemlock tree that dominates a canopy and sprinkles the forest floor beneath it with its seeds to perpetuate itself. "If this one big tree dies, the whole character of the forest is going to change," says Woodall. "Its dead wood is falling down and creates habitat for different tree species." Maybe red oaks or sugar maples replace them.

When the mix of tree species changes and some become less and some more prevalent, it might impact the ecosystem balance in the forest, which, in turn, affects how the forest sequesters carbon. "All these little changes add up and can have serious implications for the forest," says Woodall. "Even though humans that walk through the forest may not notice it."

On a sunny day in autumn 2020, Woodall walked out of his office, and, a hundred yards away, he entered a centuries-old forest. After a while he noticed a woman staring at a grove of big eastern hemlock trees. It would have taken a couple of people to encompass just one of them.

"Oh, this is so beautiful," she said.

"Yeah," Woodall replied. "But do you really know these trees?"

The woman looked confused.

"They are all slowly dying."

"What?"

"Look at the crown, the needles are turning brown and dropping from an adelgid attack. They will all probably die in our lifetime."

The woman didn't know what to say.

"She was just awestruck," Woodall says, recalling the day. "People don't think that trees die like that from forces coming out of the forest."

THE TREES CAN'T KEEP PACE WITH CLIMATE CHANGE

To a certain extent, this is all a classic example of climate-induced migration: Trees at the southern edge of their ranges are dying, whereas trees at the northern edge expand. The problem is that they don't migrate fast enough to keep pace with climate change, as several studies have shown in eastern North America. One study found that species migrate with, on average, less than half the velocity of climate change.[12] Christopher Woodall could see this in his data, too.

Climate change is encroaching on their habitats quicker than they can establish new ones. Biologists call it "colonization credit"[13] when outliers on the northern edge of the trees' range have yet to colonize areas that climate change has made available to them. It can't be explained as a simple lack of vigor among the trees in that area. In fact, these shifts ought to occur everywhere—if only slowly. But they don't. Some tree species are beginning to move, climbing up into the Alps,[14] migrating north along the French Atlantic coast,[15] or settling farther north in the northern US. However, this is not happening to all the trees, nor is it happening everywhere. Something must be preventing them from moving.

Suspicion soon turns to humans. Humans have taken ownership of the land, covering it with crop fields, settlements, and roads. We have divided up and destroyed habitats, planting forests of species of trees that don't really belong there. That's how Jonathan Lenoir explained why species on land—unlike sea-dwelling creatures—can't keep pace with climate change.

However, tree species also keep each other in check. Canadian biologists recently discovered why forests in the temperate zone found it so difficult to encroach on regions of boreal coniferous forest in the northeastern US. They call it the "priority effect"[16]: The first species that colonizes a certain area can shape it to make success almost impossible for other species. Spruce or pine forests can control light, space, and nutrients, and can prevent sugar maple or oak seedlings from gaining a foothold, even though these species are actually better adapted to the new, warmer climate. The long-established trees cover the ground with their needles and dead wood, creating a thick, acidic, dry layer.[17] Even if the climate no longer suits them, many tree species can hang on at the southerly edge of their range for decades—and fight off new arrivals from the south. However, they are no longer able to reproduce. Their populations are destined to disappear; regeneration is no longer possible. This is what biologists call "extinction debt."[18]

That's why some biologists recommend at least considering assisted migration, a highly controversial method of transplanting tree species from their old habitat to a climatically more suitable habitat. This could be within the prevalent range of a species, such as when a drought-resistant genotype from the south is being brought to the northern edge of the species distribution. Or, it can go beyond the prevalent range of a species, when species can't migrate fast enough because of barriers like towns or fields or mountains.[19]

In May 2018, Christopher Woodall and other forest experts from New Hampshire and Vermont decided to give it a try. They took thousands of seedlings from oak, cherry, and chestnut species, among others from tree nurseries all over the US, and drove three hours north to a place near the Canadian border. There, they planted them on a few hundred acres in the coarse, loamy soils of frigid spodosols within the Northeastern Highlands ecoregion.

The eight tree species were chosen because species distribution models predicted they might thrive better a bit farther north of their range in a warming climate. It didn't take long to find out: In the first year of their experiment, the region experienced a severe and unusual drought. The newcomers from the south survived.

But more than three years after the start of the project,[20] Woodall has to admit that the plan hasn't been as easy to implement as it initially seemed. One challenge has been the deer and moose that browse in the region, nibbling at the trees. He and his colleagues had to protect their plants by surrounding them with cones.

The other problem is the nature of climate change: Even if it gets warmer and maybe drier in a particular region in the long term, it cannot be taken for granted that species from the south that are better adapted to these conditions will survive. "Because all it takes is one horrible cold wave to kill everything we planted," says Woodall. And this is one of the perfidious twists of climate change: It promotes such cold spikes.[21] To make matters worse, the snowpack vanishes in the north and the trees are robbed of their warm winter coat; suddenly, they are fully exposed to winter's freezing onslaught. So far, there are a few species proving most adaptable: The northern red oak, the red spruce, and the eastern white pine have been very comfortable in their new home.

Let's assume we can help tree species move on. Still, then, a question remains: What will replace them—along with their manifold ecosystem services—when they've gone? David Moore, we know, is looking for an answer regarding sugar maples. Currently, he has put his syrup company on hold, having gone back full-time to university to study more systematically and comprehensively other tree species' syrup-making potential. His "niche syrup project," as he calls it, ranges from developing the most effective techniques for processing sap into syrup for each species to determining the nutritional and phytochemical properties of saps—not to mention the likelihood that consumers would be willing to buy them.

During the third week of November 2021, more than a hundred students, staff, and faculty members filed into an empty office on the campus of University of New Hampshire in Durham. All they knew was that they'd be participating in some kind of tree syrup study. They were each handed fifteen dollars to bid on eight-ounce bottles of syrup.

They were given two syrups to taste, without being told that one was sugar maple syrup, the other birch. When tasting the birch syrup, people showed "strong reactions," Shadi Atallah, the supervisor of the experiment, told me. Some were curious about the unfamiliar taste, while others were unhappy when sipping the less-sweet syrup, which left a slightly metallic aftertaste, as some described it.

At first, it turned out that the participants were much more willing to pay for sugar maple syrup. But after being told about the benefits of syrup tree diversification, more opted for birch syrup than before, somewhat closing the gap. Atallah takes this as a hopeful sign that education programs and eco-labeling might help to diversify the syrup market in New England. He and his team are still analyzing the results, including where the study participants grew up, to find out more about possible culturally influenced preferences. He also wants to expand the study with syrups from additional tree species. "It's worth it to continue the study," he says. "The initial results are encouraging."

However, David Moore isn't kidding himself. Ultimately, he considers his alternative syrups as novelties. "I don't think any of these other species I do research on are going to replace or even come close to replacing sugar maples," he admits. And after six years of running a birch syrup company—and many more years rigorously researching the matter—he would know.

"Nothing's going to compete with them," he affirms—not even if climate change, if left to continue unmitigated, significantly cuts their yield, as scientists predict. It will be a loss, pure and simple, and a future a little less sweet.

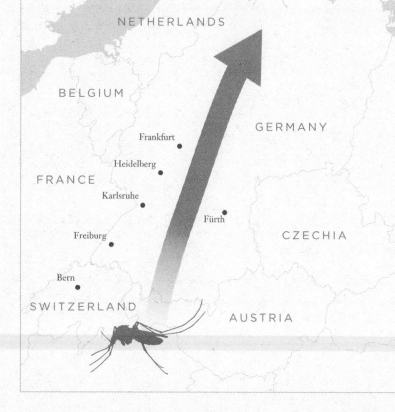

The Insects Advance

THE PROCESSION CONTINUES NORTH
Hamburg, 2019

It's not just trees that migrate. The insects that live in and off them do, too. Take the mountain pine beetle, probably the most destructive insect of all western forests in North America. Since the start of a huge outbreak in the '90s, it has been responsible for a loss of more than 25 billion cubic feet of timber.[1] Outbreaks of this species are not uncommon in North America. But climate change has favored them in the recent past, according to a 2021 Canadian Forest Service analysis, "on a scale unprecedented in recorded history in North America," which has "resulted in widespread range extensions and host shifts." It goes on, more bleakly: "Expansive tree mortality caused by this outbreak has turned the forested region in British Columbia from a carbon sink to a net carbon source, potentially aggravating the effects of climate change."[2]

The quarter-inch-long, purple-black-colored insect is feeding itself through the tissues of pine species, primarily lodgepole pines, which expand into the Yukon and the Northwestern Territories. But what keeps the mountain pine beetle in check is not only the range of its main host tree, but unfavorable climatic conditions—mainly cold winters.

With climate change, their range limits are expanding into higher elevations and latitudes, as they now have a better chance to overwinter.[3] And by conquering an area with a formerly unsuitable climate, they can do even more damage. Biologists found that *Dendroctonus ponderosae Hopkins* reproduced the best where its host trees have so far not been affected by frequent beetle outbreaks.[4]

A couple of years ago, mountain pine beetles from British Columbia even breached the geoclimatic barrier of the Rocky Mountains in a northeast direction and made it into the Canadian province of Alberta. But how is this possible, given that mountain pine beetles are not the best fliers and most of the time they move below the forest canopy, only up to 150 feet from the brood tree? The answer can be found in the air: In rare instances, turbulences can carry a beetle even hundreds of miles away.

They made it to Alberta, then, to overcome another hurdle that had prevented them from further expansion; they shifted their host to jack pine, which "has led to the entire Canadian boreal forest [becoming] at risk of invasion by this beetle."[5] For British Columbia alone, the long-term costs are estimated to add up to almost $50 billion through 2054.[6]

Other beetles are expanding their horizons, too: After the turn of the millennium, people noticed outbreaks of southern pine beetles in areas far north of their known distribution range (they are native to the forests of Central America, Mexico, and southeastern US). Huge outbreaks have been noticed in New Jersey and Ohio (2001), Maryland (2005), New York (2014), and Connecticut (2015). Recently the beetles have even reached New England.[7]

The destructive force of these little creatures affects not only the economy but our own health: When millions of coniferous trees disappear because of pine beetles, we lose their power to filter the air and remove pollutants such as sulfur dioxide, ozone, and particulate matter.[8]

Even worse are other migrating insects that can have serious direct effects on our health. These eight- and six-legged migrants include some you'd rather never meet. Mosquitoes from the subtropics and tropics are considered to be the most dangerous animals on the planet—they are responsible for around 800,000 deaths a year (not even humans, capable of war and murder, have managed that many[9]). But of the more than 3,500 species of mosquito, only a handful give epidemiologists in the temperate zone cause for concern. These species of mosquito manage to conquer new regions far from where they originate and can transmit all kinds of harmful viruses and other pathogens, including dengue fever, Zika virus, and yellow fever. Two species of mosquito in particular stand out for their expansionism: the Asian tiger mosquito (*Aedes albopictus*) and the yellow fever mosquito (*Aedes aegypti*). The name of this genus comes from ancient Greek and means "unpleasant," which seems like something of an understatement, since these creatures are capable of unleashing plagues across nations on a biblical scale.

HOW ONE BIOLOGIST HOPES TO HALT THE ADVANCE OF THE WORLD'S MOST DANGEROUS ANIMAL IN GERMANY

Melm, Ludwigshafen, Germany, August 2, 2019

The phone rang. It wasn't a number Norbert Becker recognized, but the area code rang a bell. A woman had discovered an unusual-looking mosquito in her apartment. Becker asked her to take a photo and send it to him. A few minutes later, when the image appeared on the screen of his smartphone, Becker, the former director of the German Mosquito Control Association (GMCA), spotted the black-and-white patterning on the creature's thorax, abdomen, and legs. He noticed the white vertical stripes on the back of its body and the white distal phalanx on its back legs. There was no doubt about it: it was an Asian tiger mosquito.

There was nothing special about this in and of itself. On an average day, Becker's phone would ring every ten minutes with someone purporting to have found *Aedes albopictus* or some other exotic mosquito in the Rhine Valley. But there was something about the call that day that perplexed him: It came from nearby.

The tiger mosquito had made it all the way to Becker's hometown, to Ludwigshafen, about 50 miles south of Frankfurt. It had sought him out, its nemesis.

Becker fostered a kind of love-hate relationship with the Asian tiger mosquito. The first time he had encountered one was over thirty-five years before, in Wuhan of all places, the epicenter of the COVID-19 pandemic. He was there on behalf of the University of Heidelberg, ridding the rice paddies that surrounded the city of *Anopheles sinensis*, which was spreading at an alarming rate, transmitting malaria. During the process of counting and identification, Becker came across another species of mosquito, finding hundreds in small clay pots. These, he learned, were tiger mosquitoes.

In Becker's view, there is no mosquito more beautiful than the tiger. With his tanned, striking face and his thick silver hair, Becker looks much younger than his seventy-one years. Hoping to rear tiger mosquitoes for experiments at the University of Heidelberg, Becker has been known to put his hand into glass boxes containing the mosquitoes and let hundreds of female mosquitoes spend several minutes sucking his blood. He even gave one of his daughters the middle name "Aedes" in honor of the tiger mosquito.

Nevertheless, Becker has made it his mission to prevent these exotic mosquitoes, which can grow up to a third of an inch in length, from spreading in Germany. Tiger mosquitoes are the main vector for chikungunya virus and can carry as many as twenty-one other viruses, including dengue fever and Zika virus. "The mosquito's here now and for me there's a kind of satisfaction in wiping them out," Becker explains at a café in Ludwigshafen, Germany, one day in July 2020. There is a singsong quality to his Palatine accent; he laughs and puts down his coffee cup before lowering his voice: "They're tough rivals, though."

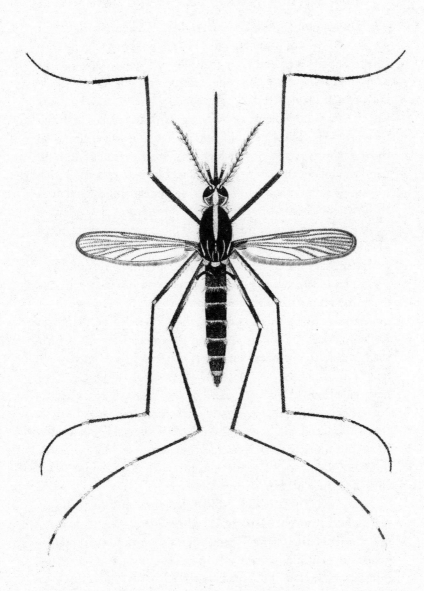

Mosquitos, like this Aedes albopictus, *are vectors for deadly diseases such as dengue fever and Zika. At specific temperatures, such as 64.4°F in the case of Zika, viruses replicate inside the mosquito's body.*

TIGER MOSQUITOES ARE COLONIZING GERMANY

Tiger mosquitoes are skilled survivalists. The females lay their eggs in old tires or potted plants. They can traverse whole continents in this way, and their embryos can survive for months or even years, even in times of extreme drought. Global trade and population growth allow these mosquitoes to travel all over the planet, finding plenty of blood to drink and places in which to reproduce. The larvae do not all hatch at once, doing so instead at a staggered pace, which prevents a period of drought from wiping out the entire population. The female doesn't stick to a single watery location when laying her eggs; she distributes them wherever she can: a few in a watering can, a couple in a bucket, a few in a swimming pool. This increases the chances of some of her offspring surviving.

In the beginning, Becker was only fighting native mosquitoes. The GMCA, a nonprofit organization, works with hundreds of municipalities to protect them from plagues of mosquitoes. Since then, however, the Upper Rhine Plain has become a gateway for tropical and subtropical insects, which catch a ride to Germany on trains or trucks on Highway A5 from Italy and find the conditions surprisingly favorable. "We have an almost Mediterranean climate," Becker remarks. In 2020, Becker retired after nearly forty years in post as head of the GMCA and, along with three dozen others, he is now able to devote himself entirely to tackling the tiger mosquito. It's much needed, too, since the mosquitoes are spreading in no time at all in southern Germany.

But it won't stop there. Moritz Kraemer and his team of researchers from the Department of Zoology at the University of Oxford[10] claim that new species of mosquito and their pathogens will continue to spread toward northern Europe. This international team of scientists has used statistical methods to link the spread of mosquito species to climate modeling and population development. The results showed that the Alps long kept the mosquitoes' spread under control, but now that this barrier has been removed, Asian tiger mosquitoes are moving north at a rate of ninety miles a year. In the next thirty years, they may well be found nationwide across Germany and France.[11]

In the US, the expansion of *Aedes albopictus* is a little bit slower, but nonetheless they are constantly moving northward at a rate of almost 40 miles a year. They were first discovered in Memphis and Houston in the '80s and then expanded most rapidly into the eastern part of the country, where they have already reached southern New England and New York. Within the next thirty years, they are expected to establish themselves across the northern US.

The spread of warmth-loving yellow fever mosquitoes (*Aedes aegypti*), which can spread yellow fever and are a key vector of dengue fever, will take some time yet—at least in Europe, where they will only make it as far as southern Italy and Turkey within the next few decades.[12] However, in the US (as well as in China), these mosquitos are gaining ground rapidly with a velocity of more than 150 miles per year, especially along the East Coast and in Texas. They might reach Chicago (and Shanghai) by the middle of the century.

Both species differ in terms of their ecology. Subtropical tiger mosquitoes prefer to make their homes in suburbs and gardens in rural areas, which offer bodies of water in which they can lay their eggs, as well as humans, hares, dogs, and cows on which to feed; tropical yellow fever mosquitoes prefer towns and cities. They prefer to drink warm human blood and reproduce in clean water tanks.[13]

According to the Oxford researchers, in the next five to fifteen years, yellow fever mosquitoes and Asian tiger mosquitoes may increasingly exhaust their ecological niches.[14] From 2020–30 onward, they may expand their reach, primarily because of climate change. Europe, China, and the US lie right on the doorstep of the region where the mosquitoes are able to survive. The two species may then establish themselves in areas where the human population is growing at a disproportionate rate: the cities of Europe, southern China, and the US. "If we don't take measures to limit the current rate of global warming, the mosquitoes' habitat may eventually extend across several cities," Kraemer explains. "There will be a large number of people who are vulnerable to infection."

A warmer and damper Europe and North America will see the world's largest increase in people contracting diseases spread by Asian tiger mosquitoes and yellow fever mosquitoes for the first time. Numbers could reach 450 million by the year 2080. And places that see human beings exposed to new pathogens for the very first time pose the greatest risk in terms of potential epidemics and severe symptoms.

All of this goes some way to explaining Norbert Becker's passion for stopping the tiger mosquito. Following the call in August 2019, he and a colleague traveled to western Ludwigshafen, south of Frankfurt, and visited the woman's house, where they placed a pill containing proteins for the spore-forming bacterium *Bacillus thuringiensis israelensis* (BTI) in the rain barrel in the garden. This bacterium originates in dead mosquito larvae and forms protein crystals that kill mosquito larvae when they ingest them. But Becker already knew it wouldn't be enough.

In fact, over the course of the summer, Becker received dozens more calls from the same area—the tiger mosquito had well and truly settled in. Becker knew he had to come up with something; he couldn't count on winter to solve the problem on its own. The tiger mosquito had successfully overwintered in the Black Forest just a few years before.[15]

The following spring, Becker took the precaution of assembling a team, including his niece and one of his students from Heidelberg. In May of 2020, the team marched out to the new housing development and distributed leaflets to more than 1,800 households, warning them not to leave water standing out in their gardens unnecessarily. Residents also received BTI pills, which they were to add to swimming pools, rain barrels, and trays under plant pots if they had the slightest suspicion that the mosquitoes were present. The first call came three days in. Tiger mosquitoes had successfully overwintered in Ludwigshafen.

NEW CONNECTIONS FORGED, OLD ONES INTERRUPTED

The COVID-19 crisis has shown us how problematic our interactions with nature are. We have spread across the world and pushed wild animals and plants into the last scraps of refuge that remain. And we even venture into these, tearing mongooses, snakes, and bats from their habitats and using them to (supposedly) boost our virility or add a touch of the wild to our dinner tables.[16] We are so close to wildlife that pathogens have few problems when it comes to switching hosts and spreading to humans.

Climate change is exacerbating this situation by turning the tables: We no longer have to go out into the natural world; the natural world is coming to us. Animals and plants are leaving their last tropical zones of refuge in their droves because these are becoming too hot and dry. They are fleeing into the surrounding villages and fields and, in time, working their way up to our latitudes.

Since animals and plants are moving toward the poles at different paces, this results in first meetings between species that have never encountered each other before. There is an increased risk of pathogens jumping from animal to animal, leading to the development of new viruses, according to the predictions of US ecologists.[17] "The redistribution of species increases the risk of completely new diseases," says Gretta Pecl, director of the Centre for Marine Socioecology at the University of Tasmania. "So many new connections are being forged, and the old ones, which once kept ecosystems and their pathogens under control, are being interrupted; pretty much everything is at stake."

There are two groups of animals spreading toward the poles particularly quickly, and they will be able to transmit diseases to humans. The first are bats, which cover huge distances and, due to the size of their populations, make excellent hosts for viruses of all kinds.[18] "If flight does allow bats to undergo more rapid range shifts than other mammals, we expect they should drive the majority of novel cross-species viral transmission and likely bring zoonotic viruses into new regions," ecologists claim. Scientists are already speculating about a direct link between the COVID-19 pandemic and climate change. Climate change has changed vegetation in the southern Chinese province of Yunnan, attracting forty new species of bat, which carried one hundred new coronaviruses into the region, according to a study from February 2021. It may have been this that encouraged the transmission of SARS-CoV-2 to humans.[19]

The second group is mosquitoes. As vectors, they are only able to indirectly transmit viruses from host to host, but with the help of humans, they can gain new ground even faster than bats. Scientists across the world are predicting the biggest increase in disease transmissions by tiger mosquitoes if the climate warms by 3.4°C (6.1°F) by the end of the century. In theory, in the next thirty years, almost half a billion more people could encounter mosquitoes that transmit diseases such as yellow fever, Zika virus, dengue fever, and chikungunya. This could rise to as many as one billion by 2080.[20]

"IS THIS A SCAM?"

On a muggy day in July 2020, Hanna Becker, Patricia Hipp, and Sophie Langentepe-Kong are going door-to-door in the district of Ludwigshafen in western Germany, all wearing the same azure blue T-shirts emblazoned with a mosquito logo. They joke that they're like the mosquito police.

"We're here about the anti–tiger mosquito campaign again," they say whenever someone opens their front door.

Since May, they have been interviewing residents, asking whether they have been bitten and examining their gardens for potential breeding grounds, which they then treat with BTI, a naturally occurring repellent. They repeat this procedure every three to four weeks. But not all the residents are so happy to invite them into their gardens. A local Facebook group has already warned of three young women going around knocking on doors in the area. "Is this a scam?" one resident wrote. But another reassured them: "They're legit!"

Everything seems to be in order in the first few gardens. The flowerpots, trays, and buckets are all dry, there's no trace of tiger mosquitoes. But then the women spot a swimming pool. "Do you change the water regularly?" Patricia Hipp asks a portly man with fair hair and stubble.

"Every four or five days," he replies meekly.

Hipp approaches the pool. "There's something in here," she says. "The edges are teeming with them."

Hanna Becker bends over the pool and spots mosquito larvae squirming about in the water. "Do you see these here?" she asks the man. "Better let the water go, as soon as possible."

"I need to kill them first," says Patricia Hipp, casually squirting some BTI in a yellow spray bottle into the pool.

But they're not tiger mosquitoes; the proboscis is not dark enough, and the shape of the squirming larvae is wrong, too. "They're just normal mosquitoes," says Hanna Becker.

There's time for one last warning, and then the three young women venture back out into the street, where Norbert Becker joins them. He has picked up a box from the GMCA office in Speyer; it arrived that lunchtime as a special consignment from Italy. Inside are tiger mosquitoes, which Becker hopes to release in the thousands in the area surrounding the housing development. "I've got the mozzies in the car!" he says happily. His three colleagues register the news without batting an eyelid.

A MATTER OF DENSITY

Some epidemiologists warn against giving in to panic as far as tiger mosquitoes are concerned. "Yes, these mosquitoes are on their way, and they can, in theory, transmit diseases," says Christina Frank from the Department for Infection Epidemiology at the Robert Koch Institute in Berlin. "But the risk of that may well be contained in the foreseeable future."

Franck wants to counter the idea that Europeans living in close proximity to these mosquitoes have the same risk of infection as people in tropical endemic countries. "The risk of outbreaks or severe endemic incidents is universally high in the places where mosquito density is especially high." And that is not yet the case in Europe or the US.

This is precisely where Norbert Becker wants to start: He wants to reduce the Asian tiger mosquito's capacity for acting as a vector in Germany. As he trudges through the housing development in western Ludwigshafen, there is the droning sound of lawn mowers, there are palm trees in one front yard, and a weak breeze brings a welcome gust of cool air. Becker pops the trunk to his matte-gray Mercedes E300de Hybrid and pulls out one of thirteen white plastic containers, each teeming with a thousand tiger mosquitoes.

Becker carries the container over to a traffic island covered with grass; two girls sitting in the shade of the entrance to a row of terraced houses eye him skeptically. He carefully removes the elastic band from the container and lifts the netting. A small cloud rises into the air, but most of the mosquitoes fall to the ground, forming a small black pile, like a pile of swept-up ash. Slowly, the pile grows smaller as more and more tiger mosquitoes whizz up into the air. There's a buzzing in our ears. But Becker simply stands there, unmoving, shaking the last mosquitoes free. He knows that these mosquitoes can't bite: They are all males—and they've been sterilized.

His colleagues at a special laboratory in Bologna have reproduced millions of mosquito eggs, sifted out the male larvae and then irradiated them, exposing them to a dose of 1.9 Gray, a unit of ionizing radiation dose, for nineteen minutes. They are Becker's secret weapon, his Trojan horse. If these mosquitoes outpace enough of the wild males, the wild females will lay almost exclusively unfertilized eggs in the area's flowerpots, bird baths, and watering cans.

The "sterile insect technique" (SIT) is nothing new. Becker's colleagues tested the method out over forty years ago, in an attempt to decimate the Rhine mosquito in Mainz. They were unsuccessful—there were simply too many of them. It was practically one sterilized mosquito to a billion wild mosquitoes. By contrast, numbers of tiger mosquitoes are still manageable. "The earlier we start tackling them and the smaller the population is, the more likely it is that we'll be able to get them under control," says Becker.

Multiple factors would have to occur together to prompt endemic disease in countries like Germany and the US. There would have to be someone returning from travel abroad who, for instance, was carrying Zika virus in their blood, and this person would then have to be bitten by a tiger mosquito or a yellow fever mosquito, which then went on to bite another person. That person would have to become infected and, in order for the virus to spread further, they would also have to be bitten by one of the two mosquito species looking for a new victim. And so on. Until recently, experts were still suggesting that the duration of the transmission season in the temperate zone would only allow for individual transmissions or only a few cases.

But is the risk of a serious outbreak not such a distant possibility, after all? A glimpse over the Alps sows the seeds of doubt.

CHIKUNGUNYA AND DENGUE ARRIVE IN EUROPE
The Adriatic coast, Italy, 2007

Castiglione di Cervia is a village of two thousand citizens, 6 miles south of Ravenna. On Sundays, its older inhabitants make the pilgrimage to the Holy Church of Antonio Abate under the shade of the pine trees. When the young people stroll off to school, they have to cross one of the two bridges that straddle the small River Savio, its waters looping slowly back and forth, with neighboring Castiglione di Ravenna nestled on its opposite bank.

The two-story houses in both settlements are surrounded by little gardens, dotted with flowers in pots. Just a few years ago, there would have been water in the trays underneath the pots, as well as in buckets and rain barrels. No one would have thought they might pose a problem. In fact, these conditions were a prerequisite for the plague that descended in 2007 and made the area famous across the world.

So what was the trigger? In mid-June, a man arrived in the village to visit his family, as epidemiologists later discovered. He had just returned from a holiday in Kerala, India, and he had a fever. At the time, India had seen more than a million people fall ill with a disease[21] that most of the inhabitants of Castiglione di Cervia had never heard of before that year: Chikungunya. The name comes from one of the east African languages of Tanzania and means, essentially, "the one who bends up."[22]

Around ten days after the man's visit, his cousin started to feel ill. Then more villagers fell sick. People both young and old were suffering from high fevers, exhaustion, and skin rashes. Some sufferers saw their joints swell up so badly that the slightest movement became unbearable. "At one point, I just couldn't stand up anymore, or get out of the car," Antonio Ciano, a retiree from the region, told *The New York Times*.[23] "I fell over. I thought, *Okay, my time's up. I'm going to die.*"

By mid-August, over a hundred residents were sick, but doctors simply could not make sense of it. All kinds of theories went around the village: The river was to blame, the government, immigrants.

At one stage, almost every household saw one of its number fall sick. But the curious thing was that family members didn't seem to be infecting each other. Suspicions turned to insects. Were sand flies to blame?

Workers at the health authority in Ravenna interrupted their Ferragosto—Italy's near-sacred summer holiday—to contact epidemiologists, send off blood samples, and set traps. When they emptied these little black containers filled with wooden sticks, they were shocked: There were no sand flies to be seen. Instead, they found tiger mosquitoes, and plenty of them.

Before 1990, there had not been any tiger mosquitoes in Italy. That year, however, a ship transporting car tires from the US docked in Genoa. Inside the tires were puddles of water. And inside the puddles were hosts of black mosquito eggs, each just 0.02 inches long.

The first tiger mosquitoes to emerge in the Italian port city appeared in a school classroom.[24] As the years went by, this invasive species spread across Italy, reaching Rome, and continuing all the way down to Sicily. In 2006, residents of Castiglione di Cervia also spotted the little insects with the black-and-white patterning.

They were just an annoyance at first. But in 2007, a situation emerged in which they were able to reveal the full extent of their skill in transmitting disease. After a particularly mild winter, the mosquitoes began hatching out of their eggs in flowerpot saucers, buckets, and puddles on April 15, much earlier than usual.[25] In doing so, they emerged at just the right time to bite the man who had recently returned from India and pass on his blood, which contained chikungunya, a 70-nanometer-long virus of the family *Togaviridae*.

The health authority informed the National Health Institute in Ravenna of a febrile illness that was displaying remarkably similar symptoms to an outbreak that had occurred two years previously on La Réunion, a French island in the Indian Ocean. A third of the population—then 770,000—had fallen ill with chikungunya virus and over 200 had died.[26]

Following laboratory analysis in late August 2007, the cases at both sites on the Adriatic were confirmed to be chikungunya.[27] It was the first outbreak of this tropical illness in modern-day Europe.

Insect control swarmed into gardens across both villages, spraying insecticide and emptying stagnant water from buckets, trays, and water fountains to dry out the mosquitoes' breeding grounds. By early September, the outbreak had been brought under control. Yet almost three hundred Italians had been infected. Three who had already been ill died; many of those who recovered suffered with arthritis prompted by the illness.

THE FIRST CASES OF DENGUE AND ZIKA IN EUROPE

In 2010, France contracted the virus: tiger mosquitoes had transmitted chikungunya in the south of the country. But that wasn't the end of it—new cases emerged in 2014 and 2017. And in 2017, the virus broke out in Italy once again. During one particularly dry summer, the mosquitoes infected over three hundred Italians in and around Rome, as well as in Calabria.[28]

In August 2010, an elderly man from Nice began to complain of fever, muscle aches, and fatigue. He also had pain behind the eyeballs, which worsened whenever he moved his eyes. The symptoms were typical of dengue fever, which had flared up time and again in the tropics since the '80s, representing "a major cause of illness and death" in the region, according to the World Health Organization.[29]

Before August 2010, only individual cases had emerged in Europe, cases of travelers who had brought the virus back with them from the tropics, without passing it on to anyone else. Yet the old man had not left France for weeks. He had, however, been visited by friends from the West Indies.

The man recovered in a few days, but, a couple of days later, an eighty-year-old man arrived at a hospital presenting with the same symptoms. He lived in the same neighborhood as the first dengue case—and so, too, it turned out, did the tiger mosquitoes.[30]

Tiger mosquitoes are also suspected in the case of the first autochthonous transmission of Zika virus, which broke out in southern France in August 2017.[31]

These diseases are clearly striking closer and closer to home, which begs the question of whether similar outbreaks are a possibility in the temperate climates of northern Europe and the US. "They are," says biogeographer Carl Beierkuhnlein from the University of Bayreuth. "We're already finding that, time and again, summers are presenting the right conditions for transmission."

As described, two things are required for infection: aggressive tiger mosquitoes, which are happy to stick their proboscis into one person after another, and hot weather lasting for several months, in order for the viruses in the insects' bodies to be at a functional temperature and be capable of developing inside the bodies of other hosts. "Climate change is extending these windows of time and making local outbreaks possible," says Beierkuhnlein. "Not mass outbreaks, but perhaps a few hundred cases."

Take Freiburg, Germany, where tiger mosquitoes have been established for a couple of years. In some summer months, temperatures in the area are as high as in southern France, and, from time to time, travelers return to the city from the tropics, bringing dengue fever with them. The fact that these individual factors have yet to come together to cause a catastrophe is simply down to luck.

The situation in the US is not much different. "Given the recent expansion of *Aedes albopictus* in the northeastern United States and the presence of major urban centers that serve as frequent entry points for travelers from endemic countries, there is an increasing threat from these arboviruses in the region," states a study published in 2021.[32] Under "optimal circumstances," the tiger mosquito could support localized transmission of the virus. And this has already happened: There have been locally acquired cases of dengue fever in Florida, Texas, and Hawaii. And in September 2013, a fifty-year-old man in Long Island, New York, was hospitalized with high fever and the typical symptoms for dengue; it turned out to be the first case of the disease in New York State.[33]

In late 2015, the world celebrated a breakthrough in the fight against dengue fever. The first, if controversial, vaccine against dengue arrived on the market.[34] The search for a vaccine against Zika and chikungunya is continuing at full speed.[35] Some believe that it's no coincidence that pharmaceutical companies are upping the tempo now, as insects and their pathogens are migrating out of the tropics and into European towns and villages. Until recently, pharmaceutical companies had little interest in developing vaccines against tropical viruses, since, Beierkuhnlein speculates, dengue and other diseases have been restricted to poorer, tropical countries for almost a hundred years. "Now that these diseases are advancing into the Global North, a market is emerging," he says.

But headlines crowing about new vaccines hide the fact that countries like Germany and the US are anything but well prepared for an invasion of tiger mosquitoes. "Politicians are just accepting it, like it's fate that we won't be able to keep it under control," says Beierkuhnlein. "They only act once the horse has bolted."

But there are things that could be done. We could monitor imports of goods, or keep a better watch over tiger mosquitoes and their movements. Beierkuhnlein has developed an app for general practitioners that shows which regions are exhibiting conditions favorable to the transmission of various tropical diseases at any given time. If a patient comes in to the practice with vague symptoms that point to chikungunya or dengue without having traveled recently, the doctor can check whether the patient might have been infected by a tiger mosquito.

Early warning systems of this kind could prevent cases from going unnoticed and allow local outbreaks to be stopped in time. But it will only work if the number of tiger mosquitoes remains within certain parameters.

TIGER MOSQUITOES CAN'T STAND THE HEAT IN THE TROPICS

To keep mosquito numbers under control, we would essentially have to stop climate change tomorrow. This would prevent them from advancing farther north. Global warming may actually help other parts of the world to rid themselves of tropical mosquitoes, as these areas become too hot for them. The malaria mosquito *Anopheles gambiae*, for example, is best able to transmit malaria at temperatures of 25°C (77°F). In 2018 alone, 228 million people contracted malaria and over 400,000 died.[36]

Thanks to its vector, the *Anopheles* mosquito, malaria is already spreading north to places where people have no resistance to the disease, while the incidence of disease in regions such as Central and South America is dropping.

The subtropical Asian tiger mosquito may also increasingly move away from the equator, according to the findings of US biologists and geographers.[37] In the event of more severe warming, regions like Southeast Asia, West Africa, and the Caribbean will benefit from temperatures rising higher than the tiger mosquito and its pathogens are able to tolerate (the optimum being 26°C, or 79°F).[38]

The bad news is that the areas the tigers free up may well be colonized by yellow fever mosquitoes, which famously prefer higher temperatures and have already spread extensively south of the Sahara. They are best able to transmit their arboviruses at temperatures of 29°C (48°F). This is already becoming apparent, with a rapid rise in cases of dengue, chikungunya, and Zika.[39]

It poses an enormous challenge for Africa's health systems, as they must orient themselves predominantly toward treating malaria. And larger outbreaks of tropical diseases are entirely uncharted waters for health systems in northern climates.

In Ludwigshafen, Norbert Becker and his young team are releasing the last few sterile tiger mosquitoes. One mother pulls her children indoors; you can never be too careful. But these mosquitoes aren't dangerous—quite the opposite. "We hope to use these to eradicate the population completely," says Becker. In the control cases, the number of viable eggs was massively reduced. In 2020, residents of the Ludwigshafen district reported two more sightings.

The odds are good that Becker will be able to get these bloodsuckers under control—at least in his neighborhood. Getting a handle on them across a whole country, however, is an ambition he dispensed with long ago.

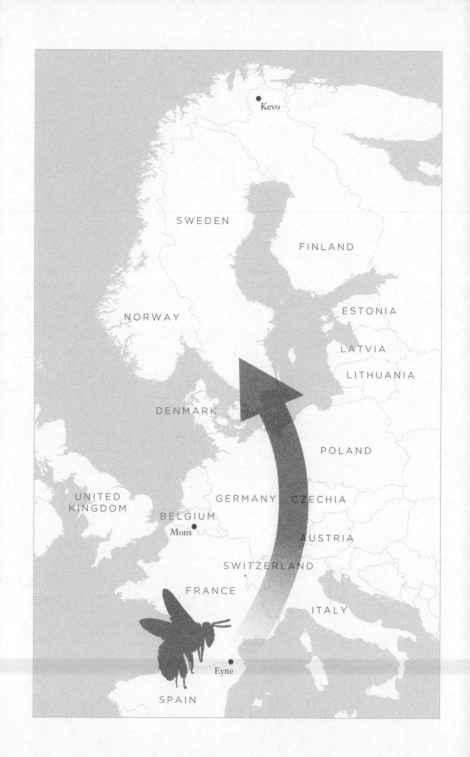

The Bumblebee Paradox

To put the image of hordes of bloodthirsty tropical insects into perspective, it's important to remember that no mosquito, tick, or caterpillar acts with malice—they simply make use of every opportunity that climate change and globalization offer them. Ultimately, it's we, humans, who alter the boundaries of their habitats, luring them into closer contact with us.

And by no means do all insects migrating northward cause harm. Many are extremely useful and contribute to species diversity. Several entomologists have expressed to me the fact that, without the hundreds of insect species from North Africa and Asia, the world of creepy-crawlies would be even more impoverished than it already has been made by intensive agriculture. Some claim that climate change is having a positive effect on insect populations, that the reasons for their decline lie elsewhere—namely, in fields, which farmers spray with pesticides and dig right up to the edges of roads and forests, due to EU acreage-dependent agricultural subsidies.

THE BUMBLEBEE KING MAKES A DISCOVERY

Even Pierre Rasmont subscribed to the theory above. Director of the Zoological Institute at the University of Mons in Belgium, Rasmont has been working on bumblebees since the '80s. He loves these peaceable creatures, known to many zoologists as "flying teddy bears" for their fluffy coats. Rasmont can tell species apart just from seeing one fly past. His colleagues know him as "the Bumblebee King."

For a long time, Rasmont couldn't imagine climate change having a tangible effect on humans and animals. What were one or two degrees in the course of a hundred years?

In summer 2003, he changed his mind. Vast areas of Europe were in the grip of a heatwave—the most severe heatwave ever recorded. So many elderly people died in Paris that funeral homes were bursting at the seams and a refrigerator intended for foodstuffs in a supermarket in Rungis had to be repurposed for the morgue.[1]

Rasmont had spent that summer—and every summer beforehand—in the Pyrenees. He and a student hoped to examine the genetic diversity of the bumblebees in the region and compare it to that of other mountainous regions in Europe. The mountain village of Eyne in the Pyrenees seemed like the perfect place to start: The wildflower meadows on the high plateau shimmer with white, yellow, and blue and are normally humming with mountain bumblebees, steppe bumblebees, and coastal bumblebees. Almost half of the sixty-nine species of bumblebee present in Europe can be found here on a single site.

In 2003, however, Rasmont presented a very different picture: The dried-out scrub had only a fraction of the usual number of insects. Scientists found themselves leaving their butterfly nets hanging up for much of the time, instead of swishing them through the swards of grass as was more usual. It was clear to Rasmont that something "worrying" and "significant" was happening, and it was all quite unexpectedly sudden.

Two weeks later, he flew to Kevo in Finland, one of Europe's most northerly regions, a mountainous area considered to be an El Dorado of bumblebees. Rasmont had once caught four hundred bumblebees there in the space of twenty minutes. But here, too, something wasn't right. The temperature had climbed to 34°C (93°F)—an incredible 9 degrees higher than had ever been recorded. Dried mosses and lichens crunched and crumbled to dust beneath his walking boots. The bumblebees were nowhere to be seen. "That's when I realized that heat waves can just wipe bumblebees out," Rasmont explains.

Rasmont, a specialist in wild bees, is sitting on a wooden bench in a meadow outside the Zoological Institute at the University of Mons. A warm autumn breeze blows linden leaves into his beret, which is sitting faceup on the wooden table in front of him. Two COVID-19 cases were reported at the institute just this morning, so we fled the great brick building for the outdoors. Rasmont waves over colleagues or students from his research group as they walk past every few minutes; one places a little pack of pills for Rasmont's sick turtle on the table, another brings a tin of muffins, one explains in detail what they are working on at the moment. Rasmont wouldn't be able to manage his workload without them; he has too many questions that he still hopes to answer, primarily that of the future of bumblebees in northern climates. His professorship resembles a bumblebee colony in a way, with numerous worker bees (students), and a couple of drones (colleagues), who do the bulk of the legwork for the benefit of the queen (Rasmont himself).

Rasmont's interest in bumblebees developed out of his love of the mountains. There are over 250 species buzzing about on the planet today, and they emerged over thirty-four million years ago in the mountains of Asia, during the transition from the Eocene to the Oligocene, a period in which our planet was rapidly cooling. This cold phase may have pushed the bumblebees down from the then young Himalayan mountains, spreading from these cooler pastures all the way across the world.[2] Today, bumblebees are the sole pollinators in many mountain regions, surviving the cool temperatures thanks to their dense, fuzzy hair. For Rasmont, it's been a lucky break; in order to get to the bumblebees, he explains, he "simply had to" hike the French Pyrenees, the Scottish Highlands, and the Köroğlu Mountains in Turkey.

Since 2003, however, the bumblebees' refugia have changed. Places that have been subjected to heat and drought have seen the density of their bumblebee populations collapse by one or two orders of magnitude depending on the site. All of them are relatively untouched, not developed or cultivated or treated with pesticides. Rasmont discovered that bumblebees were four times less common in hot, dry years than in cold, wet years. He named this phenomenon "Bumblebees Scarcity Syndrome."

Previously, climate change had played a somewhat obscure part in discussions surrounding insect decline. But now, Rasmont and his colleague Stéphanie Iserbyt were asking whether the consequences of global warming may well have been evident for some time—in the form of extreme events such as storms, drought, and heatwaves. "Is it possible that such events that are not accounted for by global statistics act on the fate of the bumblebee's faunas?" the pair asked in a journal article published in 2012.[3]

Rasmont still didn't know exactly what had killed the bumblebees. At first, he assumed that they had starved or died of thirst because the flowering plants had all dried up or burned, and the water had simply evaporated. Yet this kind of event would have seen all species across these regions collapse equally—and this wasn't the case. Another possibility was that the heat had directly killed the bees. Rasmont had already observed bumblebees simply dropping dead and falling out of the sky on particularly hot days.

To test this hypothesis, Rasmont and his protégé Baptiste Martinet placed one bumblebee after another inside a heated incubator normally used for rearing reptile eggs. He showed me such a device in Mons: It resembled a microwave, fitted with a temperature display. Inside, male bumblebees were exposed to temperatures of 104°F,[4] until they toppled backward and twitched their legs—an indicator of heat shock. For each species, Rasmont and Martinet simulated how long an insect's body would be able to tolerate the heat: Polar species could stand only twenty minutes, many species from the temperate zone were able to withstand two or three hours, and ubiquitous species such as the buff-tailed bumblebee were able to endure as much as twelve hours.

The incubator experiment demonstrated that bumblebee species react very differently to heat. Surprisingly, however, Rasmont determined that different populations of the same species behaved in exactly the same way, whether they were in the North Cape or Spain. This suggested that they were not able to adapt to the hotter, drier conditions at the southernmost edge of their range. Heat resistance was fixed in every species.[5] Rasmont was alarmed by this discovery; it was not good news for bumblebees. If climate change makes our summers hotter and drier, this will increasingly exceed the heat tolerance of most species, particularly at the southernmost edge of their distribution. Sooner or later, these species will disappear from these regions.

Hoping to get to the bottom of these concerns, Rasmont collaborated on an EU project together with institutes from sixteen countries, gathering more than four million units of distribution data on wild bees and bumblebees in Europe and compiling them in a database.

A macroecologist from North America was pursuing the same issue in parallel. Years earlier, Jeremy Kerr from the University of Ottawa had proved that butterflies in Canada were shifting their habitats northward. Now he wanted to test to see whether the same applied to bumblebees. In 2010, he encountered Pierre Rasmont at a conference in Pisa, Italy.

"I've just come from northern Alberta, where it's fifteen degrees warmer than usual," Kerr explained to Rasmont. "And I couldn't find a single bumblebee of any kind."

"I've just come from North Finland," Rasmont replied, "And I observed exactly the same thing."

The two men had both spent many years of research in the Arctic, which had given them a keen eye for the effects of climate change. While still in Pisa, they planned out a study they would embark upon together, examining how bumblebees on both continents were reacting to climate change. They built a database for the project, feeding it with over 420,000 observations of a total of sixty-seven species of bumblebee from Europe and North America, spanning a period of over one hundred years. The oldest observations came from museums. Next, they checked where different species of bumblebee had lived at different points in time; once for the years spanning 1901–74, and a second time for the years that followed, during which global warming had begun to pick up speed. The results showed a clear pattern: Bumblebee species were losing their most southerly habitats and had so far retreated north by an average of 190 miles.[6]

The losses were particularly heavy in Spain and Mexico, where whole bumblebee populations were disappearing or fleeing up into the mountains.

By contrast, on the northerly edge of their distribution, the bees were hardly able to conquer any new ground, unlike birds and butterflies, which have shifted their European habitats northward by an average of 23 and 70 miles, respectively, and yet are still lagging behind the pace of climate change.[7] Kerr has compared the bumblebees' area of distribution to a carpet being rolled up from its southern tip.

WHO'S TO BLAME?

Who's to blame for the bees' retreat? To find out, Kerr and Rasmont compared the range shifts with long-term data on maximum temperatures, pesticide use, and landscape reconstruction. A direct correlation emerged for only one factor: climate change.

Rasmont and Kerr wanted to publish their findings in the renowned journal *Nature*, but after a year and three rounds of reviews, the magazine rejected the paper. "There's a paradigm issue in the US," Rasmont complains. Many ecologists, he claims, are convinced that pesticides, habitat destruction, and the fragmenting landscape play a huge role in insect decline—which they undoubtedly do. "The problem is that this leads them to refuse to accept that bumblebees may *also* be victims of climate change."

After the study was published in *Science*,[8] some of these critics piped up. Yes, climate change had hit bumblebee populations in their southernmost range, but the real causes of the problem were still unclear. It could be, they said, a result of parasites, which are better able to spread and attack bumblebees due to climate change.[9] Or perhaps global warming had simply dealt the fatal blow once they had already been weakened by intensive land use. Take Cullum's bumblebee, for instance: These little creatures, with their black-and-pale-yellow fuzz, have lost a majority of their habitats since 2010, as more and more wildflower meadows are cultivated for agricultural land.[10] At some point, the only place where they continued to thrive was the French Massif Central. And when a heatwave hit the region in 2003, with temperatures more than 18°F above normal, *Bombus cullumanus* disappeared from its final patch of habitat.[11]

Rasmont and Kerr had found a pattern, but they had yet to discover the mechanism behind it. Kerr arranged a follow-up study that same year. Together with his colleague Peter Soroye, he looked back over where and when, since the beginning of the last century, hot spells and droughts had cracked the historic heat threshold of various bumblebee species. This, they discovered, correlated with the sites where the insects were disappearing.[12] In other words, the probability of bumblebees occupying a certain area decreased in every location where climate change had pushed species beyond their limits of thermal tolerance. Their habitat in Europe had shrunk by 19 percent, and by almost half in North America.

Kerr and some of his colleagues believe the bees are just the beginning. The more frequently that heatwaves and droughts increase in severity, the more the tolerance thresholds of other insects will be exceeded and the more they will lose parts of their southernmost areas of distribution—not due to a gradual rise in temperatures but rather because of extreme events occurring with greater frequency and severity.[13] "Most species don't even experience 'climate' per se," Kerr explains. "They experience the weather."

WHERE HAVE ALL THE BUMBLEBEES GONE?

It's a sunny late summer's day and I'm on my way to Friedeburg, a village in central Germany that sits on a hill on the banks of the River Saale. There, I meet Oliver Schweiger from the Helmholtz Centre for Environmental Research in Halle. Schweiger is a macroecologist with shoulder-length hair, and he leads me into a permanent observation area; data from this site was included in the study by Jeremy Kerr and Peter Soroye. An owl flies out of a tree. Ravens and buzzards circle in the air, sending their cries echoing through the valley. It is a special place: Plants and animals, for which these latitudes might otherwise have been too cold or too wet, have been able to settle here in the rain shadow of the Harz mountains. These include bee-eaters, which build their nests in the hollows in the loess rock—farther north than ever before. However, three years ago, it was too dry even for this area of low precipitation. "In 2018, we had just ten inches of rain—half the long-term average," says Schweiger. "These are the kinds of numbers seen on the steppes."

The situation improved only marginally in the two years that followed; there was still too little rainfall. And the effects were everywhere to see: The meadows, which would usually have been grazed by sheep, were withered and dry; there were just a few sparse flowers blooming, yarrow and viper's bugloss, thyme and sage. On the phone, Schweiger reassured me that we would still find all kinds of bumblebees at this time of year. But, apart from a few honeybees, there are no pollinators to be seen. Perhaps it's too windy. Perhaps the vegetation is already too dry. Analysis of previous years may provide greater insight, but these are still pending. But Schweiger does not want to exclude the fact that some bumblebee species are beginning their climate-related retreat earlier than expected in some of Germany's warmer regions. Of thirty species, around half are now on the red list, making them particularly vulnerable to heat waves, drought, and storms. This won't be good news for the farmer I spot busy plowing a neighboring field of oilseed rape.

As we come to the end of our loop around the hill, Schweiger stops for a moment in a meadow scattered with horse dung and crouches down. "Here's one!" An orange-and-black carder bee moves busily from one flower to the next. Soon, it sets its diagonal course for the blue sky above us and flies off. *Bombus pascuorum*, the most common species of bumblebee in Europe. Yet even this little bee may have disappeared from Germany by the end of the century, according to modeling. Much like the buff-tailed bumblebee, it may be able to seek refuge in the cooler climes of Northern Europe. Yet the majority of bumblebee species seem to be held back by an invisible barrier, like a nightmare where you're being chased but your feet won't seem to budge. "We don't know why it is," says Schweiger.

Remember the BioShift project led by Jonathan Lenoir, who discovered that sea-dwelling creatures are always hot on the heels of their isotherms, while land-dwelling creatures fall way behind? This, Lenoir explained, occurs because the loss of habitat and the fragmenting landscape prevent them from spreading north. Strangely, this does not seem to apply to bumblebees: They do not persist exclusively in heavily managed and fragmented landscapes; they can also be found in relatively untouched surroundings. We can only speculate as to what actually prevents them from moving. Many bumblebee species do not tend to swarm out. And even if they are in the position to scatter, they must first establish populations in the new locations, from which sufficient queens can strike forth into the new area. But there may well be no suitable flowering plants in the area, or there may be a period of a few warm years followed by a cold one, which leads the new colony to freeze.

POLLINATION TAKES MORE THAN JUST HONEYBEES

In spite of everything, a few species are succeeding in making the move. The buff-tailed bumblebee isn't choosy about food and has successfully covered large distances in its journey north. Pierre Rasmont assumes this has something to do with the bee's nighttime behavior. After sunset, the queens fly for miles, marking a trail with scent to attract males.[14] "And this could be the moment when they expand," Rasmont speculates.

The buff-tailed bumblebee has already crossed into the Arctic Circle. A few years ago, Rasmont discovered these jacks-of-all-trades 500 miles north of their previous limit of range, on an embankment in the mountains of Norway. Their huge colonies made the most of all kinds of flowering plants in the area and displaced the small colonies of Arctic bumblebees. "It's quite clear that *Bombus terrestris* will replace other Arctic fauna in the near future," says Rasmont.

However, Rasmont is not inclined to complain about the buff-tailed bumblebee; after all, they are of great help to humans. Millions of hand-reared swarms of bumblebees pollinate tomatoes, apples, cherries, kiwis, peaches, and all kinds of berries, in greenhouses, fields, and orchards. Three quarters of all crops humans depend on for food are themselves dependent on pollinators.[15] Some people see the buff-tailed bumblebee as an alternative to what is currently the most commercially significant pollinator: the European honeybee, *Apis mellifera*. Honeybee populations have been in decline for a few years. An incredible fifty billion bees died in the winter of 2018–19 alone; the bees had been intended for pollinating almond trees in California[16] and may have died due to a combination of

The carder bee (top) is Europe's most prevalent species of bumblebee. Its northward permanent migration may leave none left in central Europe by the end of this century. It's a similar story for the buff-tailed bumblebee (bottom), which have recently been found in the mountains of Norway, 500 miles north of their previous limit of range.

pesticides and the introduction of the *Varroa* mite.[17] However, the costs of carting millions of these creatures across the US on low-bed trucks to pollinate almond trees in California, apple trees in Washington, and blueberries in Maine are simply astronomical.[18] In experiments, bumblebees have been shown to be at least on a par with honeybees as pollinators, if not more effective, for a range of crops such as watermelons, cucumbers, and pumpkins.[19] While honeybees only fly to the flowers they can easily pollinate, bumblebees leave no flower untouched. They use their long tongues, dipping them into the nectar and pollinating the moist and sticky stigma in the carpel with pollen from other flowers. They also use a special technique known as buzz pollination. They beat their wings, achieving specific frequencies and shaking the stamens of certain flowers that are harder to reach, such as those of tomatoes, potatoes, and blueberries. They set to work in wind and rain, at twilight, and even at 50°F. It explains why bumblebees are the main pollinators of some crops in many regions, regardless of whether honeybees are used.[20] Places where bumblebees are active often boast crops of better quality, and harvests increase. "Bumblebees are the best pollinators we have in wild landscapes and the most effective pollinators for crops like tomatoes, squash, and berries," says Peter Soroye. "Our results show that we face a future with far fewer bumblebees and much less diversity, both in the outdoors and on our plates."

Even today, many crops are underpollinated.[21] Environmental economists in the US have found that yields are particularly low in regions experiencing a dearth of wild bees.[22] This may be due to farmers turning the bees' habitats into cornfields and spraying them with pesticides, or the increase in temperatures and drought may be driving them away. Rasmont has used computer simulations to calculate that the buff-tailed bumblebee will have to relinquish large areas of its region of distribution in Europe by the end of the century because these places will become too hot. For now, their current range reaches as far as the Sahara in the south. In more unfavorable scenarios, this line may rise as high as Paris and Mainz; the more favorable scenarios would still see this rise as far north as Madrid and Rome. "Then agriculture in southern Europe will be practically impossible," says Rasmont.[23]

But the disappearance of bumblebees doesn't just have consequences for farmers: 85 percent of wild plants are only able to reproduce with the aid of pollinators. Without them, whole landscapes would no longer flower. And without wild plants, many animals would no longer be able to find the berries they depend on for food, ecosystems would collapse, and people, too, would be affected, because the ground would no longer take up water, leading to landslides and flooding.

As the global species carousel sends our pollinators northward, it also brings us all kinds of pests from the south, which compete with humans for food. Fungi, insects, snails, and rodents already obliterate half of the world's harvests.[24]

Bioscientists at the University of Oxford have examined hundreds of pests in the northern hemisphere and calculated that they have been migrating toward the Arctic at a rate of almost 2 miles per year for the last half century.[25] Rising temperatures mean that not only will there be more of these pests but they will also be better able to reproduce and multiply and the temperature will accelerate their metabolisms. All of this may lead to significant losses of our harvests in the future, particularly in the most productive regions of the world: France, the US, and China.[26]

Farmers will have to get their thinking caps on if they want to manage the pests and also feed an additional two billion human beings, which the United Nations expects to see by the middle of this century.[27] One solution might be new crop rotations. Experts estimate that more pesticides may help on intensively cultivated land. Today, farmers spend $30 billion per year on insecticides, herbicides, and fungicides. The dosage increases each year as the pests develop resistance. But what's really crazy is that pesticides don't just kill pests; they also kill all kinds of pollinators, as well as the predators that feed on the pests: rodents, birds, spiders, wasps, flies, and bugs. Pesticide use can see even the most harmless of species become pests, as they start to spread without any competition to keep them in check.[28] It is true that some countries in the north are benefiting from new fish species, pollinators, and other groups. However, in general, it remains the case that the more we cause the Earth to warm up, the more complicated and the more unpredictable it will become to feed the world's populations.

"THE VERY OPPOSITE OF OUR POLICY"

On November 14, 2016, Pierre Rasmont stepped into a glass building in Brussels. Thirteen years had passed since his shocking discovery in the Pyrenees, and Rasmont had now gathered sufficient findings to be able to predict what would become of Europe's bumblebees in an ever-warmer world. He aimed to prepare the members of an EU parliamentary committee of experts for the fuzzy little creatures' migration. By the end of the century, he explained, over three quarters of species could lose the majority of their habitat, and a third would go extinct if climate change continued at its current pace. Bumblebees would only retain a few sites in the south, which boast a particular microclimate. Sites of this kind, such as the Forêt de la Sainte-Baume near Marseille, were particularly worth protecting, he explained: The two-thousand-year-old beech forest in southern France lies on a mountainside, which shields it from the heat of the surrounding area. The forest affords numerous species of bumblebee refuge and a place to keep cool, while the mountain's western flank is covered in dry, Mediterranean vegetation. At the same time, Rasmont explained, new species such as carpenter bees would advance from the south. A couple of these chunky little juggernauts have already established themselves in Germany.

Indeed, they are among the regular guests that have made their home in my parents' Bavarian garden over the past couple of years, overwintering and nesting in a woodpile. They look like oversized bumblebees with jet-black bodies and blue, iridescent wings. They can sting but have generally left us in peace so far. There are some bumblebee species that are benefiting from climate change, such as the clay bumblebee, *Bombus argillaceus*. These enormous bumblebees may colonize all of Germany and France in the coming decades.[29] "We ought not to stop them," Rasmont explained on his visit to Brussels, recommending that the parliamentarians go so far as to support wild bees in their efforts to migrate to cooler habitats. If entire species are at risk of disappearing, assisted migration could be the right option.[30]

Upon hearing this, the committee chair fixed her eye on Rasmont. A few months previously, the EU had approved a list of unwanted, invasive species.[31] This offered profiles of what now amounts to sixty-six plant and animal species introduced to Europe from other continents, spreading rapidly and compromising local ecosystems. It included muskrats, raccoons, and bullfrogs. And now here was this Belgian zoology professor, telling her they ought to be welcoming the invaders from the south? It was the very opposite of the policy that Europe had been adopting for the past two years, she replied.

But Rasmont wasn't denying that exotic species from other continents could pose problems. In most cases, they cause little disruption, or even benefit the local ecosystem, but now and then a new species will hit like a meteorite. In Canada, the zebra mussel originating in Russia and Ukraine has practically sterilized some aquatic ecosystems. The North American gray squirrel threatens the domestic squirrel, and the North American comb jelly has almost wiped out anchovies and sprats in the Black Sea.[32] Insects such as the Asian tiger mosquito could pose real problems for our health systems in the future. What Rasmont was criticizing is the stance taken against any form of migration—just like the belief that one can seal off an entire continent. He sees parallels with the debate on refugees.

In fact, a xenophobic tone of sorts resonates throughout the introduction to the EU booklet on the "Black List"; some ecologists even speak of "bioxenophobia."[33] "Ecological barriers like oceans and mountain ranges have allowed ecosystems to evolve independently, so that the species within them are adapted to each other and interact in a delicate balance. Moving species across those barriers can severely disrupt this balance and may even change these ecosystems entirely."

At the heart of this lies a mindset that has prevailed in ecology to this day: the notion that species exist in a stable balance of sorts and that human beings arrived and disrupted the equilibrium. What matters now, according to this mentality, is restoring the former conditions so that animals and plants can recover. "Nothing is in equilibrium!" Rasmont tells me, outraged. "Thanks to climate change, we are living in a state of complete disequilibrium." Yet as species attempt to find solutions to this imbalance by changing their habitats, conservationists and politicians seek to push them into fixed areas, "protecting them" within nature reserves and within the borders of specific countries.

"We have to abandon this idea of nature as static," Rasmont says. Millions of animals and plants have begun to migrate across the globe under their own steam. They leave their reserves, cross borders, and migrate from one continent to another. But many politicians still don't seem to have accepted this. The belief that we will be able to control the influx of new species persists in many ministries and congressional and civil service offices.[34]

Rasmont recommends reconsidering this limited focus on invasive species, because what occurs in the majority of cases is actually a natural process of migration—as in the case of the scarlet dragonfly and the wasp spider, which are spreading in Germany. And even in the case of introduced species, this often simply preempts their imminent migration, itself prompted by climate change—as in the case of the ring-necked parakeet from India[35,36] or the African sacred ibis, an enormous bird with white plumage, a black head, and a hooked bill, which hunters in Brittany are shooting in the thousands because it supposedly threatens tern colonies. The Egyptian otter is currently losing its habitat in Libya and Egypt as a result of climate change, and Rasmont goes so far as to envisage its assisted migration to the Var in southern France, which boasts a similar climate. "But at the moment it's simply impossible to ask for anything like this," he says. "Politics rejects anything that might look like an 'invasion.'"

Our ecosystems could experience severe problems in the future if not for the species that are arriving here from the south. We need them because they fill in the gaps left by species that have migrated away. "We have to accept this and organize, not try to prevent it," says Rasmont, because, after all, the stream of species moving north can no longer be stopped. They are being driven on by a force stronger than any border.

"They are coming," says Rasmont.

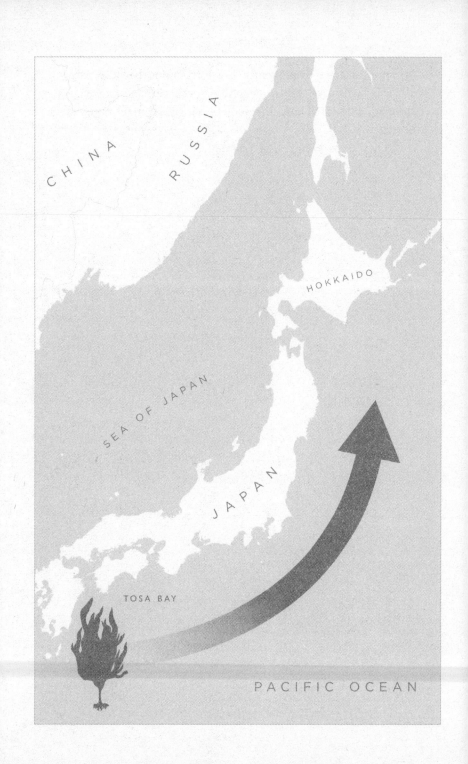

Cultural Assets under Threat: Japanese Kelp

The bumblebee king is not alone. Biologists from across the world have reported repeatedly coming up against brick walls when they try to tell politicians about the phenomenon of species migration. There is a real reluctance to recognize it at all. Where does this ignorance come from? Perhaps the problem is not considered sufficiently pressing, especially since climate change throws up so many issues in need of solutions. Perhaps the answers posited simply seem too complex and overwhelming.

But maybe there is a deeper-rooted reason for it all. The migration of species is a reflection on us and gives us pause. By streaming toward the poles, scaling mountains, and crossing oceans, animals and plants reveal how ridiculous our borders are, along with our illusion of control. The same can be said for the belief that we can degrade nature to a certain extent and outsource the wilderness to a few remaining conservation areas. According to the latest predictions, by the end of the century tens of thousands of species could cross national borders and leave some countries altogether.[1] In doing so, animals and plants not only remind us of our questionable contact with nature but also bring our minds back, relentlessly, to climate change, which they embody with their exodus across the globe, making it all the harder for us to ignore the crisis we are in.

All this explains why there is no international agreement on or even recognition of this phenomenon, and why countries are responding individually to the migration of species, instead of tackling the issue together. They do so on a case-by-case basis, protecting, pursuing, or ignoring these species depending on whether they are of benefit to a country or are considered under threat there, whether said species impinges on local ecosystems or is hardly noticeable.[2] "We can't leave biological diversity to chance; it's crucial for humankind's survival," says marine biologist Gretta Pecl.

The story of the kelp forests off the coast of Japan reveals just how futile such efforts can be.

Tosa Bay, Japan, 1997

It was just the odd tale at first, but eventually they came together to form a consistent picture: Japanese fisherman were returning home with less and less fish in their nets. Instead, they reported a puzzling phenomenon: The kelp forests just off the coast were disappearing quickly.

Kelp forests are composed of macroalgae and resemble real-life forests, but underwater. Diving into their world, you find yourself floating through underwater woodland. Their adhesion organs lie on the seabed, the stalks rising straight up, growing to almost 200 feet, up to their thallus lobes, which stand in the water column thanks to their swim bladders, transforming the light that falls just below the water's surface into energy. Kelp forests sprawl across several levels, forming a canopy that, when the sunlight breaks through, throws shadows on the brown-golden algae below, making these cold-loving symbiotic communities as bewitching and magical as coral reefs. Kelp forests do not play host to as many species as coral reefs, but they provide refuge for more endemic species that only exist in specific areas. One such species is the magical Glauert's sea dragon, which lives south of Australia.

Kelp forests are of national importance in Japan, where people have gone so far as to erect shrines to this brown seaweed.[3] Rich in iodine and other trace elements, kelp is added to almost every dish. Those gathering the seaweed take to boats to cut the ends of the thallus lobes, which may be used for soup (kombu) or for rolling sushi (nori).[4] Kelp forests also provide refuge for commercially important species of fish, gastropods, and crabs, which find themselves on Japanese dinner tables and help keep the local fishing industry afloat.

And so, Japan was particularly alarmed by the loss of its kelp forests. It wasn't an entirely new phenomenon; in the 1980s, a monitoring program carried out by the ministry for the environment revealed some decline, which was attributed to turbulence in the ocean and sea urchins. However, in 1997, this phenomenon began to accelerate dramatically: Tosa Bay lost almost one square mile of its kelp forests. Of the once-dominant brown seaweed *Ecklonia cava*, only a few shredded and munched scraps remained, and a few years later, many of the rocks were completely denuded or covered in little more than a scant layer of seaweed. The Japanese have a name for this phenomenon: *isoyake*. The term brings together two words meaning "rocky coast" and "burned."

But what was the reason behind the kelp forests' collapse? At first, the coastal prefectures speculated about sediment deposits on the sea floor, typhoons, or runoff from nuclear power plants.[5] But since the kelp forests were collapsing almost everywhere—until almost half had disappeared completely—the cause, they determined, could not be restricted to one area: It must be everywhere, like the warming of the oceans, which Japan is particularly affected by. The waters off southwestern Japan have warmed at a rate of 0.3°C (0.54°F) per decade since the 1980s. In Tosa Bay, winter temperatures had risen to 1.7°C (3.06°F) warmer than they had been three decades before.[6] This poses a problem indirectly for rootless brown seaweed because it requires nutrients to grow, but these nutrients decrease as the water warms. It also affects the seaweed directly because the plants wither when temperatures climb too high. This is what happened in

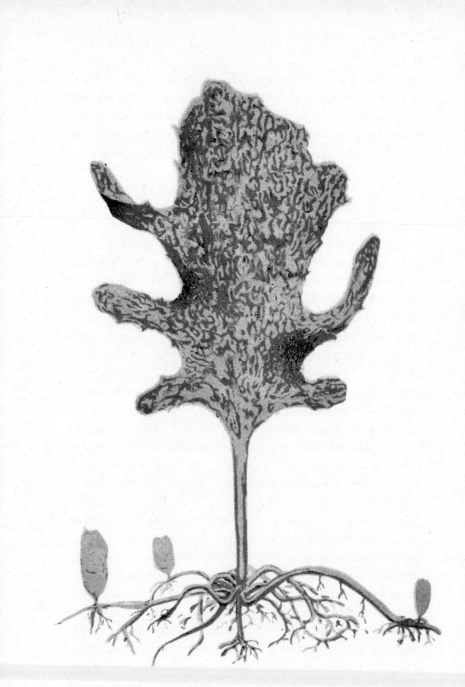

The decline of brown algae (Ecklonia cava) can be attributed to rising seawater temperatures, which attracts plant-eating tropical fish and forces the algae to retreat to colder waters.

western Australia: In 2011, an extreme marine heatwave wiped out the kelp forests along a 60-mile stretch of coastline in one fell swoop. The seaweed lost the northerly extent of its range of distribution and is increasingly being pushed toward Australia's southern continental shelf.[7] But was it really that simple?

Institute of Marine Science, Sydney, 2011

In November 2011, marine ecologist Yohei Nakamura flew to Australia to take part in a workshop on the tropicalization of temperate marine ecosystems. The workshop was given by Adriana Verges, another marine ecologist, who was working at the University of New South Wales and investigating a new phenomenon. She had established that the construction of the Suez Canal in Egypt had led all kinds of herbivorous tropical fish from the Red Sea to migrate into the Mediterranean and graze on its kelp forests. Verges wondered whether the same could have occurred naturally, in places where warm ocean currents wash plant-eating tropical fish into subtropical and temperate waters. It was nothing new per se, but climate change may now accelerate the process, such that the waters in these warm-water currents warm by two to three times the global average. Sporadic incidences of individual tropical fish could become permanent tropicalization. At the workshop in Sydney, Verges addressed marine ecologists from across the world, asking, "Is there evidence that this is already happening?"

In the years that followed, scientists set out to test Verges's theory, in regions such as eastern Australia, for example, at the edge of the kelp forests' range of distribution in the transition zone between the tropical and temperate zones in the Pacific. If anything was going to change, it would be here. Luckily, underwater cameras had been positioned on the sea floor in the area for some time. When Verges assessed the ten years of available footage, she noticed that the kelp forests were indeed constantly shrinking, until they disappeared altogether. In some locations, scientists had also placed samples of the dominant kelp species *Ecklonia radiata* on bare rocks to see what would happen. On the screen, Verges saw a school of rabbitfish nibbling on the kelp's thallus lobes as they swam past, a gray chub chewed away at the kelp, and then a collector urchin snuck up on a blade of kelp, hoping to attach itself to it and gorge itself. Evidently vulnerable, the brown seaweed was consumed within hours.

Scientists in other parts of the world made similar observations—particularly in Japan.[8] In Tosa Bay, Yohei Nakamura was able to follow how the rabbitfish *Siganus fuscescens* and the Japanese parrot fish *Calotomus japonicus* overgrazed the kelp forests as the sea warmed. Gradually, it became increasingly clear that the fish were a crucial factor in disrupting the balance of the heat-struck kelp forests and causing them to struggle to recover from subsequent heatwaves.[9]

But the rocks did not stay bare for long. Where the kelp had once been, something new emerged: corals.

CORALS IN JAPAN? CORALS IN JAPAN!

Tosa Bay, Japan

Corals, hybrids of coral polyps and alga, are the epitome of sedentary species. Corals are attached to the substrate beneath them. Unlike fish, they cannot simply swim away; if their environment deteriorates, they have to endure it. However, their offspring can seek out more favorable habitats. Once a year, corals release eggs and sperm in unison. After fertilization, larvae form in vast numbers, resembling an enormous pink cloud. These sometimes spend weeks or months floating at the ocean's surface. They follow the warm ocean currents, which carry masses of water along the east coast of the continents toward the poles, like the gulf stream in the North Atlantic, the East Australian Current in

Australia, or the Kuroshio in Japan, which is also known as the Black Current for its dark hue and carries warm water out of the tropics, as well as tropical fish and coral larvae.

The corals were once merely accidental guests who would ultimately leave again when the winter cold set in, but the situation has changed since the 1980s, when the waters off the southern coast of Japan began to warm. As a result, their numbers exploded; summers saw over a hundred coral species dwelling in Tosa Bay. A few years ago, water temperatures exceeded the threshold at which coral species can overwinter (15–18°C/59–64°F).[10] Some of them, such as the stony corals *Acropora muricata* and *Acropora latistella*, became permanently established. The corals benefited from the work carried out by herbivorous fish, which prepared the ground for them. By keeping the kelp forests low, they ensured that these would no longer overshadow the corals. Looking at it another way, coral communities offer these once-vagrant fish shelter and feeding grounds where they are able to settle permanently. When corals emerge, they usually have fish in tow. Nakamura was the first to discover proof of this change in regime. When he goes diving in the bay, he spots neon-blue damselfish and the near-circular, colorful butterfly fish swimming around the stony corals, but no kelp.

Meanwhile, in other parts of the world, reports of new coral reefs growing outside the tropics are increasing: Stony corals have been spotted in Sydney Harbor, overwintering on the city's doorstep. They have worked their way about 190 miles down the east coast of Australia, following the current, accompanied by tropical fish like the sergeant major fish *Abudefduf saxatilis* and the black butterflyfish *Chaetodon flaviostris.*[11]

Similarly, corals, tropical fish, and other warm-water species have taken over the degraded seabed off the west coast of Australia, where the kelp forests fell victim to a heatwave.

Staghorn corals and elkhorn corals have expanded out of the Caribbean, into the Gulf of Mexico, and 30 miles up the coast of Florida, emerging farther north than ever before.[12]

Around the world, corals are shifting their areas of distribution to latitudes 33 and 34 degrees north and south of the equator. Warm ocean currents are like highways, which the corals use to race toward the poles at a rate of 9 miles each year.[13]

Yet the Japanese are less enthused about their new guests, as beautiful as the corals and their friends, the tropical fish, may be, and no matter how many tourists they entice to the country—particularly since the abalone population, which made its home in the kelp forests, has also collapsed. In 1996, fishermen were still hauling 1.9 tons of these mollusks out of the waters of Tosa Bay; by 2000, hardly any remained. Langoustines and squid also disappeared; many fishermen lost their livelihoods overnight.

The residents of Japan have lived off the kelp forests for over a thousand years; nowhere else in the world has supported as many species—thirty-eight in all. This explains why they are reluctant to give up without a fight. The national fishing authority published a guide for dealing with the isoyake issue now affecting almost all coastal areas. The guide lists twenty-five techniques for stemming the collapse of the kelp forests, including releasing seaweed spores, distributing concrete blocks or stones on which the macroalga can settle, as well as protecting the area with metal spikes or grids. The most important aspect, however, is the request made of the fishermen: They are to "remove" and "reduce" herbivorous tropical fish and sea urchins.[14] In other words: Stamp them out.

But what are fishermen supposed to do with all the rabbit-fish that land in their nets? The authorities have come up with a solution for precisely this: Japanese diners, they say, should simply change their gastronomical habits and eat them. However, these exotic species must first be made appetizing; a conference paper from 2019 indicates that many people are put off by the fishes' smell and the "poisonous spines" on their dorsal fins.[15] Yet experts are doubtful about the strategy's success. "The problem is that fishermen are aging, and there are less and less of them as time goes on," Nakamura explains. "So there's a lack of manpower to implement these measures."

In the long term, Japanese kelp forests may well cease to exist even without competition from the south. The center of kelp production is located off the coast of the island of Hokkaidō, in the more northerly, cooler region of Japan—beyond the reach of rabbitfish and others. But the ocean is also warming in this region, causing the kelp forests to retreat. The kombu harvest has halved in the last three decades, while the price has doubled. This could radically alter Japanese cuisine in the long term. Scientists from the Hokkaidō Research Organization have calculated that the sea around Hokkaidō will have warmed by 10(!)°C (18°F) by the end of the century and that kelp forests could lose three quarters of their habitat, if not all four.

But future generations will not have to swear off kombu altogether, as it is already being increasingly cultivated in land-based containers filled with seawater. However, many traditional kombu producers do not consider this a viable alternative, because the quality is inferior to that of naturally grown kombu.[16]

Another matter for debate is the question of whether it is ethically justifiable to inhibit the migration of sea urchins,[17] seaweed-eating tropical fish, and reef-forming corals, no matter how much they threaten the old ecosystems. "At events, I'm often asked, 'What about species that move into another area and negatively impact the ecosystems there, be they predators or competitors?'" Lesley Hughes explains. Hughes reported the early migration in a meta-study carried out over twenty years ago. "So then I ask them, 'Do you shoot them, or do you give them a medal?' You might want to shoot them because of their negative impact, but you might want to give them a medal for adapting to climate change."

The corals and tropical fish in Tosa Bay are the best example of this. On the one hand, they are driving back the kelp forests, but on the other, they are simply trying to remain within their thermal niche, adapting to environmental changes that we have inflicted upon them.

They're heroes, in a way: They've succeeded in escaping the inferno and saved their species by finding a new, habitable spot where they are able to survive.

PART III

Exodus in
the Tropics

A Dirty Little Secret

When mosquitoes, wild bees, birds, fish, and even corals move away from the equator in droves, from Mexico and the Caribbean to the southern US, from North Africa to southern Europe, they pose new challenges for human societies. But what is happening in the tropics now that its inhabitants are migrating? Other species from hotter regions can hardly move in and fill in the gaps because it's already the hottest zone on the planet.

Residents of the tropics may have some resistance because they have adapted to the heat over millions of years. This was the British naturalist Alfred Wallace's view; almost 150 years ago, Wallace described the climate of tropical forests as possessing an "extreme equability and permanence." "Every form of vegetation has become alike adapted to its genial heat and ample moisture, which has probably changed little even throughout geological periods."[1]

But what if it hasn't? A nightmare scenario flitted before my eyes: the gradual disappearance of the most species-rich region in the world, condemning the people of the tropics to starve to death. I decided to investigate the matter and I found my first clue far back in the past.

GeoCenter Northern Bavaria, Erlangen, Germany, 2018

The walk to Wolfgang Kiessling's office is a journey through Earth's history. In the garden, there are ammonites; the steps lead up past a tableau of fossilized corals, 200 million years old, and his room on the first floor of the old building hosts a replica of an archaeopteryx.

In 2017, Kiessling, a paleoenvironmental professor, carried out a bold experiment. He examined the migration of sea creatures since their emergence 450 million years ago, using data from millions of fossils. He used statistics to calculate their previous spread and attributed this to the particular plate tectonics at the time. The next thing he did was use isotope analysis of fossils to examine how temperature fluctuations in the oceans behaved in relation to this over the course of Earth's history. Kiessling and his colleagues were surprised by the clarity of the fossil report: It clearly pointed to a changing climate.[2] Corals, mussels, and sponges had repeatedly shifted toward the poles and then moved back depending on whether the climate was warming or cooling. "The poleward shifts performed by these creatures may well have been normal in the past," says Kiessling.

He examined case studies from the past, trying to find an analogy for the present. He happened upon a period 125,000 years ago. At this time, too, the climate was warming; the global average temperature was 1°C (1.8°F) warmer than it is today. Kiessling looked at high quality fossils of corals, which showed that they had migrated at this time, too. Their range shifted as much as 500 miles toward the poles—admittedly over a period of two thousand years.[3] The tropical belt also exhibited only mild warming of less than 1°C in contrast to higher latitudes, but cnidarians nevertheless migrated away from the equator to the mid-latitudes in great numbers. The diversity of species in this area increased, while decreasing in the tropics.

This led Kiessling to conclude that if migration, not adaptation, had always been first choice when it came to reacting to climate change, it presented an enormous challenge for tropical species in the present day. These species are already having to deal with climatic conditions that haven't existed for hundreds of thousands of years. "Temperature shelter in high latitudes may not be sufficient to counteract the loss of equatorial diversity in a warming world," Kiessling explains. In other words, only a portion of the species from the tropics will be able to save themselves by escaping to higher latitudes.

I wonder if this exodus out of the tropics is imminent once again. Could it have already started?

Kiessling furrows his brow. It's possible, he says, but it's a difficult question to answer. Unlike more northern latitudes, the most species-rich regions of the world are lacking in data. Large swaths of the tropics are essentially blank spots on the research map.

This is outrageous, I think. So much is at stake, but it has hardly been researched. It's like a dirty little secret.

Kiessling leaves me with one more candidate. The first place the retreat is likely to occur is in the ocean, where it will affect the organisms that are most sensitive to heat: corals. Their gentle spread into temperate ocean could represent the first step in a mass exodus.

PACIFIC OCEAN

CORAL SEA

GREAT BARRIER REEF

AUSTRALIA

12

—

The Corals Move Out

Great Barrier Reef, 2017

Terry Hughes could hardly believe his eyes as he flew over the Great Barrier Reef in March 2017. Beneath the shadow of his small aircraft, Hughes, director of the Centre for Coral Reef Studies at James Cook University in Townsville, Australia, watched reefs roll by, one after another. But instead of the usual luminous colors, all he could see were white skeletons. Two thirds of the reef off the northeast coast of Australia had been affected by coral bleaching. Only the remaining third, to the south, was in its usual condition.

Hughes, one of the world's most renowned coral scientists, began his career in the Caribbean. When the coral reefs in that part of the world deteriorated in the early '90s, Hughes moved to Australia to research the Great Barrier Reef, the largest coral reef in the world. It was still in good condition. "I'm sort of an ecological fugitive in search of a pristine coral reef," Hughes once said in an interview.[1] But he soon learned that there was nowhere to hide from climate change. There have been four great incidences of coral bleaching on the Great Barrier Reef since 1998.

Coral reefs are unique habitats. These enormous reefs are created by invertebrate organisms, often less than half an inch in size: coral polyps, their cylindrical bodies ending in mouth-like openings, surrounded by tentacles. These colonies of cnidarians are constantly secreting calcium carbonate, and over thousands of years this creates the basic structure for their underwater backdrops. Marine biologists estimate that as many as a million plant and animal species have made their homes here, protected by the coral reef. A quarter of all animal species in the world's oceans depend on these reefs and the food and refuge they provide.

However, corals can only survive in symbiosis with minuscule algae known as zooxanthellae, which lodge themselves in the polyps' skin, where they draw off carbon dioxide. These single-cell organisms require carbon dioxide to photosynthesize. In turn, they excrete oxygen and glucose, supplying these to the coral polyps. It is these algae that lend coral reefs their luminous red, green, and blue colors.

If the water warms by more than 1°C (1.8°F) above the highest average summer temperature for a sustained period, the algae begin to produce radicals, and eventually the polyps decide to show their lodgers the door. All that remains is a white skeleton of calcium carbonate. The polyps can survive for some time without the algae, but eventually they starve.

After his surveillance flights, Terry Hughes slipped into his diving suit and surveyed 104 reefs in the immediate vicinity to get a clearer picture of the situation. The deserted skeletal reefs dominated primarily in the north, boasting a few remaining scraps of rotting coral tissue, already hogged by thread algae. The southern part of the reef had emerged practically unscathed, possibly due to the cooling effects of Cyclone Winston. Despite Hughes's dismay, this presented an exciting experiment for him. It enabled him to answer a crucial question: Just how much heat can corals survive?

Analysis showed that coral cover barely changed if the water did not rise more than 3°C (5.4°F) above the long-term average. Above this point, however, the mass die-off began. At a level of heat stress exceeding 4°C (7.2°F), coral cover receded by 40 percent; above 8°C (14°F), this increased to two thirds, and above 9°C (16°F), it was as much as 80 percent. Hughes put the critical threshold at 6°C (11°F). Above this point, he identified changes in the composition of coral species: The stress had initiated rapid selection between the different heat-resistant corals. Staghorn corals and purple hood corals—known as "Milka corals"—were particularly affected. In areas that had been hit by severe bleaching, these fast-growing, delicate-branching species shrank by as much as three quarters, after which more robust, slower-growing corals began to dominate. "Coral die-off has radically changed the composition of coral species in hundreds of reefs," says Hughes.

Coral bleaching is now rampant in all tropical waters around the world. Between 1980 and 2016, scientists from Australia, Britain, and Canada investigated bleaching in a hundred reefs across the globe and found that the likelihood of coral bleaching had increased fivefold. In 1980, the likelihood of bleaching in a region remained at an average of once every twenty-five to thirty years, but by 2016, this had risen to once every six years. The Caribbean was particularly severely hit, and the seas of the Indian Ocean and to the east of Australia have been most affected recently. "Tropical reef systems are transitioning to a new era in which the interval between recurrent bouts of coral bleaching is too short for a full recovery of mature assemblages," Hughes wrote in a sensational study published in the journal *Science*.[2]

Hughes feared how long it would be before the next bleaching would set in on the Great Barrier Reef, because the time between these bleaching events is crucial for the corals' survival. One bleaching cannot kill a reef, but each one is a blow to the resilience of the reef, especially when these come in quick succession. It's much like the way that a boxer can take a punch but eventually hits the floor when subjected to a whole barrage of blows.

In fact, it takes corals a good ten years to fill in the gaps left behind by corals that have died off. The Great Barrier Reef has not had sufficient time to recover since the bleaching of 2017, however. In early 2020, Hughes took another trip in his light aircraft over the reef, which covers 214,000 square miles off Australia's west coast, and saw the corals glimmering, snow-white, beneath the water, with single dabs of yellow and green here and there. It was no longer happening just in the north, as with the previous bleaching; this time, the south was affected, too. This kind of spread had never occurred before.

Hughes has calculated that, long term, the reef will continue to degrade until climate change is stabilized and the remaining corals transform into new, heat-tolerant reef communities—provided that they prevail over their notorious rivals, the seaweeds, which attempt to overrun the corals and even deploy chemical arsenals against them.[3] "What is currently taking place underwater is nothing less than chemical warfare," says US marine ecologist Mark Hay from the Georgia Institute of Technology in Atlanta.

A special report by the Intergovernmental Panel on Climate Change (IPCC) warns that, by the end of the century, almost all coral reefs may have disappeared entirely, even if we succeed in limiting global warming to a maximum of 2°C (3.6°F), as specified in the Paris climate treaty.[4] A limit of 1.5°C (2.7°F) would save at least 10 to 30 percent of these reefs. Yet even after three mass bleaching events on the Great Barrier Reef within the space of four years, the government of Australia has refused to commit to emitting less carbon dioxide and exporting less coal. Instead, it has made do with absurd token measures such as distributing sunscreen across the water's surface and installing giant cooling fans on individual reefs.[5]

A MASSIVE GAP IN THEIR DIETS

In May 2017, I visited the islands of Fiji to report on the effects of climate change on the Pacific Island state, which was due to lead the United Nations Climate Change Conference (COP23). It was here that I saw a coral reef for the first time. I could never have imagined the magical world that existed beneath the ocean's surface, parallel to ours. The reef was covered with corals of every color. Starfish bent their arms and revealed their little doll-like faces; clownfish with white and orange stripes swam over to us, full of curiosity; and a school of swordfish rushed past, just an arm's length from my goggles. Everywhere, there was a bustle of activity. I was captivated by the magic of it all and never wanted to return to the surface.

Coral reefs are tremendously alluring. They speak to something deep within us, a longing for rich diversity and nature. But does the loss of them have practical consequences, too? Something that goes beyond the thought that, one day, tourists from the north will no longer be able to fly halfway across the world to marvel at their beauty? "The consequences will be dire," explains Alistair Jump, a British ecologist. "Anyone who thinks that they exist independent of biodiversity is completely bonkers—utterly mistaken. The loss of biodiversity will impact everybody."

This includes coral reefs. Reefs boast whole reservoirs of medicinal resources that can be used to tackle all kinds of complaints and diseases. Nowhere else on Earth is host to such fierce competition; here, countless species are forced to fight for space, and sponges, corals, and seaweed produce chemical components that can be used to fight antibiotic-resistant bacteria, viruses, and cancer.[6] Meanwhile, pharmacists are also interested in the millions of microorganisms that live on coral polyps and beneath the silty seabed. "It's a veritable library of answers," Hay gushes. "But in most cases, we don't yet know what the questions are."

Roughly 850 million people living on the coast depend on the fish on the coral reefs, which they either eat or sell, or on revenue from tourism. It's a simple formula: If the coral reefs are happy, then so are the people who live near them. Without the reefs, the fish will eventually disappear. This was proved a few years ago by Morgan Pratchett and his team from James

Cook University, Australia. The team observed that the corals were disappearing. At first it seemed as if the fish were able to continue their lives as normal without the reefs. But when the scientists examined the situation ten years later, they found that most of the fish had vanished.[7] Without fishing, the economies of many of the smaller island nations would collapse, forcing some of their citizens to starve. According to modeling, fish production from coral reefs could shrink by a fifth by 2050. At the same time, however, the population of Pacific Islands such as Samoa and Vanuatu is growing vigorously. There's a real risk of a massive gap in the population's diets.

Many are setting their hopes on tuna fish. Tuna fish live in the open ocean and could compensate for the lack of reef-dwelling fish, to some extent. However, because of climate change, the two most significant species, the yellowfin tuna and skipjack tuna, are also migrating—eastward. They are turning their backs on the Maldives in the Indian Ocean and the Cook Islands and Fiji in the South Pacific. "They are leaving behind nations that have long made a vast proportion of their income from these fish," says Ove Hoegh-Guldberg, director of the Global Change Institute at the University of Queensland. This income is primarily from fishing license fees paid by foreign trawlers, which is practically the sole source of revenue for some island nations.[8] A fish factory providing employment for one thousand people on the island of Lovoni, part of the islands of Fiji, was forced to cut back production of albacore tuna in late 2019.[9]

Tuna fish will increasingly leave the economic zones fished exclusively by the Pacific Island nations and push out into international waters. Large trawlers from countries such as China are ready to land these fish as soon as they arrive. For the time being, they are still mostly paying fishing licenses, but the tuna's migration eastward will give them free rein. Fishing experts estimate that the Pacific Island nations could lose $60 million per year within the next thirty years.[10]

Fish is the sole source of protein for many inhabitants of the remote islands in the South Pacific. When local fishermen are unable to find the tuna fish, they frequently take their boats to fish in the coral reefs—until those fish also disappear. "Some communities get so desperate, that they're trying to pick the last fish from the reefs," says Hoegh-Guldberg. "They're driving the reefs to extinction."

A few years ago, Hoegh-Guldberg visited Indonesia, where he heard reports of coast dwellers using dynamite to hunt on the reefs, hoping to grab the last remaining fish. "Many people who live on the coast have no other choice," Hoegh-Guldberg explains. "We expect impoverished, starving people to behave differently from us and protect the world's ecological well-being, while we pump the Earth's fossil fuel reserves dry and heat up the planet, causing unsolvable problems for those who live on islands in the tropics." Experts are now recommending that island dwellers—people surrounded by the ocean—fish inland, in lakes and rivers, to source the requisite quantity of protein.[11]

SALVATION AT HIGHER LATITUDES?

So, what about the coral reefs? Will they succeed in decamping to higher latitudes and cooler oceans before they lose their heartlands on the equator? Nichole Price and her team of marine biologists at the Bigelow Laboratory for Ocean Sciences in East Boothbay, Maine, anticipated precisely this, so they took a look at an important indicator of coral reef health: procreation. They used terra-cotta tiles to calculate new coral recruitment in all tropical and subtropical waters in the last forty years for the first time.

Since 1974, hundreds of these tiles have been sunk in tropical and subtropical ocean regions to see whether new corals would establish on them. Findings showed that repopulation by corals had dropped by 82 percent. This provided proof of the crisis the corals are facing; mass die-offs and disease were causing the corals to reproduce less. However, this decline was not the same everywhere. While the number of young corals in the tropics collapsed by 85 percent, those in the subtropics increased by more than two thirds—in the north of the Gulf Coast along Texas, in Hawaii, on the south of the Great Barrier Reef, and in Egypt, all places where new coral reefs had recently been discovered.

This, Price says, offers "a glimmer of hope." At the very least, Price writes in a study for the *Marine Ecology Progress Series*,[12] high latitudes may be able to offer some coral species "an ecologically meaningful refuge against adverse conditions likely to occur in tropical seas." According to distribution modeling, in just a few years' time, the best habitats for corals will no longer lie along the equator but instead above 20 degrees of latitude.

But getting there is no picnic. The larvae need to be able to draw on their fat reserves for long enough to be able to cover the vast distances required. Next, they must find a suitable subsoil on the seabed, preferably rocks in shallow water, which receive plenty of light throughout the year. Once they have settled here, they still have to wait for their symbiotic companions, the microalgae.

Stony corals have proved that it is not an impossible task, letting themselves drift up from the stricken Great Barrier Reef toward the coast of Sydney, or from the degraded Caribbean reefs toward the coast of Fort Lauderdale, Florida. It is not just temperature that determines coral viability; it is also light and the depth of the ocean. Without light, the algae are unable to photosynthesize. Consequently, corals can survive provided that they do not stray too far from shallow waters or from the equator. The farther toward the poles they travel, the darker it is in winter, and the more acidic the waters become, as carbon dioxide dissolves better in cold water. Tropical reefs are like sandwich fillings, Price says: They can only have a future within a relatively narrow zone between the tropics and the subtropics. Something new might well emerge there, but it could take awhile. The formation of a coral reef the size of the Great Barrier Reef could take, roughly, half a million years.[13]

An Abrupt Change of Regime

Recently, an international team of ecologists and climate scientists tried for the first time to compile a timetable of when, and in what regions of the world, temperatures would grow too hot for certain species if climate change continues at its current rate. The computer simulation spat out its answer: 2074.

This suggests that, in fifty years or so, the thirty thousand land- and ocean-dwelling species the scientists selected will experience radical upheaval and a drop in their diversity, according to the study, which was published in *Nature*.[1] Of course, this number only represents an average value, which means that, in some regions, the collapse will occur later, while in others it will take place considerably earlier. Scientists have deemed the risk to be particularly high for the tropics. These regions are not only rich in species, meaning they have more to lose, but also exhibit a dearth of species that are adapted to warmer conditions. Even today, the study's authors explain, the first animals and plants are being exposed to temperatures that they have never experienced before, and this is occurring not by means of a gradual increase in temperature but in sudden heat waves, the foot soldiers of climate change.

In the past thirty years, the number and duration of heat waves has increased by 50 percent. These have been raging in the waters of the Caribbean and off Australia, accelerating the poleward migration of species[2]—as in the cases of the kelp forests in western Australia and the coral reefs in eastern Australia.

"It's not a slippery slope; it's more like a series of cliff edges, which affect different areas at different times," says the study's coauthor, Alex Pigot, from University College, London.

Scientists predict that around the middle of the century, problems present in the ocean will affect land-based systems and erode iconic ecosystems such as the rain forests of Indonesia, the Congo Basin, and the Amazon rain forest. But it may well happen earlier than expected.

I envisioned tropical species migrating en masse, fleeing up mountains and traveling away from the equator to seek out cooler climes. It was a veritable exodus, combined with a massive loss of species diversity.

But can it be true? Sure, I can imagine fish migrating out of the tropics and into higher latitudes; sea-dwelling organisms always respond much faster to temperature changes. But how in the world is the Amazon rain forest, with all its resident flora and fauna, supposed to move to a higher latitude? And what would replace it?

None of this fit my image of the majestic, evergreen Amazon rain forest, which has existed for at least fifty-five million years and continues to play host to the greatest number of species in the world. In northern temperate zones, we've almost grown used to the loss of plants and animals, safe in the knowledge that there is still a sizable refuge for a diverse number of species in other parts of the world, such as tropical rain forests and coral reefs. If these also collapse, the world will be irretrievably impoverished, as will the way we envision it. It seemed mind-boggling to me.

In my search for answers, I decided to travel to the part of the world that boasts the greatest variety of species: the mountains of the Manú National Park in Peru.

14

The Mountain Forest Begins to Climb

Kosñipata Valley, Peru, Summer 2019

I could tell what made the place so special before the plane had even landed. Seemingly endless forest extended from the lowlands up into the Andes. Meandering rivers carved out wild loops, sometimes mirroring each other, as if they were two of the same. I was overwhelmed to see so much untamed nature; my eyes were used to fixed shapes like canals, straightened rivers, and square miles of woodland.

Somewhere down in one of the valleys below, Kenneth Feeley was hiking through the Amazon rain forest in the ocher-colored garb he always wore in the tropics. An ecologist at the University of Miami, Feeley was spending the summer in Manú National Park, tree hugging. It wasn't for spiritual reasons, of course; he did so to measure the thickness of the tree trunks. His research began in southeast Peru in 2003. At altitudes of 3,100–11,150 feet, US biologists mapped fourteen plots of forest, each of them larger than a soccer field. In four years' time, they repeated the process. After the second survey in 2007, Feeley urged his doctoral supervisor, Miles Silman, to analyze the data. The team were not expecting a significant change in distribution during this brief period, but they seized the opportunity to practice for later analysis. However, they soon encountered a pattern: tree species were migrating upward, by 8–11.5 feet on average.[1] "We certainly expected something like this to emerge," Feeley explains. "But not for another fifty years."

Schefflera, *a common houseplant, is a particularly skilled climber—a crucial advantage in the age of climate change, as it can quickly scale mountains in search of cooler ground.*

Of course, trees can't simply haul the roots out of the soil and skip off up the mountain. But with a little help from gravity, wind, and forest-dwelling animals, they can distribute their seed. In the age of climate change, the direction these seeds will travel in is generally a given: If the seeds travel too far down the mountain, they will die, because this area is now too hot for them; if they travel up the mountain into cooler climes, they will survive because they are adapted to the temperatures in this area. This is how trees climb mountains. Plants of the genus *Schefflera* are particularly skilled climbers. *Schefflera* is more familiar to us as a houseplant, but in the wild it grows quickly and spreads its seeds, with some help from bird species such as the tanager, which picks up the seeds and then excretes them at higher altitudes. Trees that simply drop their seeds, however, may not be able to keep pace with climate change.

Feeley and his colleagues wanted to know whether this phenomenon was restricted to the Manú National Park. They linked up with scientists from across the world and began recording ever-larger forest plots, in Colombia, Ecuador, and northern Argentina. Today, the census comprises a total of two hundred plots. And the same phenomenon is emerging everywhere: Cold-loving tree species are moving to higher ground, where it is cooler.[2] Peru was not an anomaly.

It may sound simple, but the trees' ascent to greater heights is anything but. It's not just temperature that is crucial for their well-being; humidity and cloud cover are also important. And these factors are not changing in sync with global warming. Take humidity, for instance: In contrast to temperature, humidity rises the higher a mountain is. So, if the air warms permanently due to climate change, allowing it to absorb more water, species will have to migrate down, not up, in order to remain in an area boasting the same degree of humidity as before. US biologists established this while investigating bird species living in mountain transects, mapped a century before, in the Sierra Nevada. Most of the birds had shifted the spread of their breeding sites—but only half of them had traveled upward to do so. As global warming pushed them upward, the increasing humidity drew a smaller number of the population back down.[3]

And even if some species can climb high enough quickly enough, these new pastures may well present conditions that the species cannot survive. The forest boundary illustrates this well enough; instead of shifting upward, the boundary proved to be surprisingly stable for the two hundred plots, Feeley explains. Comparisons of old photographs, US military aerial shots, and field observations revealed that the sites where the trees stood highest fifty years ago are the same sites boasting the highest-growing trees today. Feeley speculates that this could be due to the temperatures in the tropical mountains, which fluctuate wildly from day to night. If tree seeds blow across the forest boundary into the grasslands, they will die from exposure to the cold night air. Beneath the forest canopy, however, they are protected enough to survive. "It's a catch-22," says Feeley. "You can't have a forest without the seedlings, and you can't have seedlings without the forest."

These transitional habitats, or ecotones as they are known, such as the forest boundaries, are present at different transition points in the mountain forest. For many species, the climb ends far below the summit. "The tree species that are moving up from below will be compressed at the tree line," says Feeley.

He still can't make out the shift in the rain forest just by looking. But he has been noticing a different change for a few years now. On the mountain peaks, Feeley has spotted black, charred patches that were green in previous years. New patches appear every year. "A substantial portion of the uplands has been burned off," says Feeley. These areas are now grazed by cows or used by farmers for crops. "The trees want to move up but people are pushing them farther down."

But if the forests are damaged, there are consequences for the entire food chain. Trees provide shelter for vast numbers of animals, fungi, mosses, and lichens. Many insects have adapted to live on particular trees, with birds adapting to live on those particular insects. There is an especially large population of such birds in Manú National Park; one in every ten of the world's bird species calls the region home. The birds were my reason for coming to Peru in the first place. I hoped they would tell me what was happening in the tropics.

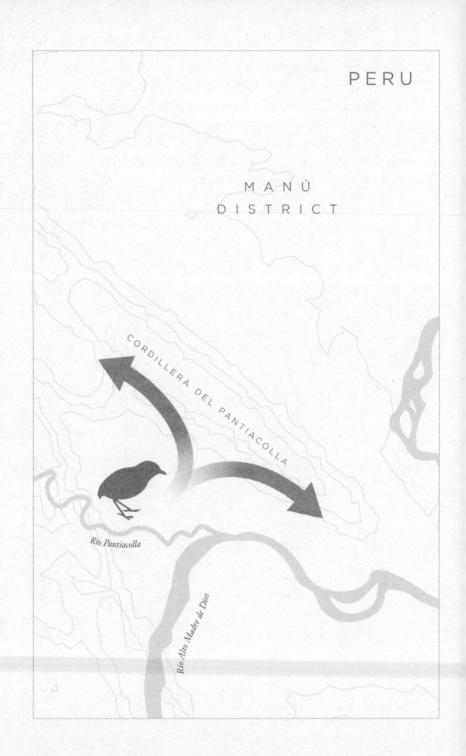

PERU

MANÚ
DISTRICT

CORDILLERA DEL PANTIACOLLA

Río Pantiacolla

Río Alto Madre de Dios

15

—

The Escalator to Extinction

Cusco, 2019

I've arranged to meet Alex Wiebe over 11,000 feet up, in Cusco, the onetime capital of the Incan Empire. Wiebe is a biology student from the Laboratory for Ornithology (the Cornell Lab) at Cornell University, Ithaca, New York, a training ground for the best ornithologists in the world. We plan to climb a remote patch of the Andean foothills in the southeast of the country. A strange phenomenon has come to light here on the Cordillera del Pantiacolla, one the science community has called "the escalator to extinction" but which has never been unequivocally documented before.

The story of this discovery begins with a swashbuckling trip taken in 1985. A couple of US ornithologists, now legends in their field, traveled to Peru to scale an "untouched" mountain and observe the local birdlife. Among them was John Fitzpatrick, now the director of the Cornell Lab, and one of Wiebe's idols. The seventy-year-old is known for starting lectures by putting his hand to his gray mustache and imitating the call of the barred owl. Fitzpatrick was the same age as Wiebe is today when he first set out for Peru to identify the birds in Manú National Park. It was a time when ornithologists let their beards grow long, swarming out to the most remote corners of the Earth to discover uncharted animal and plant populations— using shotguns when necessary (birds shot on the Pantiacolla are listed as "collected"). "Everything we know about tropical birds in Peru is down to Fitzpatrick and a few of his colleagues," says Wiebe.

Thirty years later, Fitzpatrick set out for the Cordillera del Pantiacolla once again, this time to accompany a protégé of his, Benjamin Freeman. The two men stood by the river, held up a black-and-white photo from 1985 and compared it with the bank until they found the right spot. No one had climbed the mountain in more than thirty years, and yet the biologists realized that something fundamental had changed: The birds had shifted their habitat up the mountain by an average of 220 feet.

As they climbed higher, the number of individual birds shrank along with their habitat. Eventually, once the pair had reached the top, the birds disappeared altogether. Fitzpatrick and Freeman proved that eight species were affected. The cause was barely perceptible: The air has warmed by 0.42°C (0.76°F). This was the first time they had documented entire species communities vanishing beyond the mountain's peak due to climate change.

Alex Wiebe now hopes to test this incredible discovery. If anyone's up to finding birds that might no longer exist, it's him. In July 2019, Wiebe, a baby-faced student built like a football player, took part in a "Big Day" at Manú National Park, a competition that saw fifty-eight birdwatchers given twenty-four hours on foot to find as many birds as they could. Beforehand, Wiebe had read up on what species were present in the region, where they lived, and at which time of day they sang, and picked his route accordingly. He set off early, at 2:45 AM, and pitched back up at the Los Amigos Biological Station at 10:30 PM. He had spotted 347 bird species—a world record.

Wiebe loves competition, he tells me, as we sit in the sunshine on the balcony of a restaurant in Cusco eating *pollo a la brasa*. We can hear the drones of car and scooter horns from the road, belching out clouds of soot, making it even harder to breathe. And here, at an altitude of 11,150 feet, breathing is no joke. Peru is home to 1,802 bird species; Wiebe already knows most of these, at least those in the lowlands. Now he wants to scale the tropical mountains, starting with the Cordillera del Pantiacolla.

Seated in a Mercedes Sprinter, we set off from Cusco, traveling seven hours on dirt roads down through the Andes into the lowlands in the east of the country; past slopes where Peruvians in colorful clothing drive oxen, plowing furrows in the fields; past boys leading their donkeys in the opposite direction, up hairpin bends and, once it gets dark, past motorbikes and 4x4s that prefer not to switch on their lights, or simply follow a flashlight, in order to save on gas. The settlements in the rain forest have names like "Cross," "Three Crosses," and "Holy Cross." We are spending the night in *Salvación*.

The next day, three young men from the Palatoa community, an Indigenous group that lives in the area, pick us up by a channel in a wooden boat with an outboard motor. The Palatoa community belongs to the largest Indigenous group in Peru, the Machiguenga, who mostly keep to themselves, but they're not averse to sporting national soccer team strips and carrying simple cell phones. Up from Río Alto Madre de Dios, a river as broad as a lake, we turn into the Río Pantiacolla, which wends its way through the rain forest. Then it appears before us: the Cordillera del Pantiacolla.

The mountain range is pyramid-shaped, covered with forest—or forests. At the foot of the mountain ridge 1,500 feet up, guadua bamboo, palms, and giant trees are abundant. The temperature drops a good 0.5°C (approx. 1°F) every 100 yards, and the trees dwindle. High up on the precipice, they grow stunted in the wind, and the moist mountain air cloaks them in mosses and creepers. "The vegetation determines where we find which birds," Wiebe explains. The bird communities living at the foot of the mountain are completely different from those living at the summit. Mountain ridges such as these are among the few that can boast such rich diversity in such a narrow space.

Our boat glides out by the earthy bank below the mountain, exactly where Fitzpatrick and Freeman once disembarked. Wiebe's rubber boots slap their way through the water at the edge of the bank. "You can wash here," says César, our cook and helper. But he recommends that we first use a stick to test the riverbed for stingrays and reminds us to watch out for caimans. The embankment leads up into a clearing. We camp here for two days, while Angél and Kevin, two Palatoa boys, cut us a path to the top with their machetes.

A film of perspiration coats our skin; it is so humid, our bodies feel like they're swimming away as we tackle the mountain flank—over half-crushed stems and dried palm fronds that crackle and crunch beneath our feet, through thorny branches of bamboo canes that lean over the path. The trunk of a kapok tree fans out along the ground in a triangle, a walking palm bears its roots six feet above the ground; this is supposed to help with ventilation. If we were to fell one of these trees at random and observe a cross-section of the trunk, we would instantly be able to see why the rain forest and its inhabitants face such a threat from climate change: They have no tree rings.

In the latitudes I am familiar with in Germany, trees grow at different speeds according to the season; this leads to lighter early wood and darker late wood. In the tropics, however, there are no seasons; trees can grow all year round in relatively constant temperatures. This means that birds can nest at any time of year. The rain forest has been able to lay claim to this area over millions of years, while as recently as twenty thousand years ago North America and Europe saw ice sheets pushing their way down, towering thousands of feet high. According to one theory, species in the tropics had time to grow apart from each other in all conceivable shapes and colors and occupy their own niches. They were able to settle in paradise.

The same took place on the Cordillera del Pantiacolla. We keep seeing clumps suspended in the branches of trees, like medicine balls painted black. These are termite nests. Even trogons nest here. Other birds, such as the blue-headed macaw, live in hollows in the trees. And others, like the russet-backed oropendola, live in nests that hang from the trees like bags.

In just half an hour, Wiebe has filled three pages of his notebook with acronyms, four letters that stand for the species' name, and then strikes for the number of animals spotted. He explains that they are all birds that have so far benefited from climate change. They are still able to cope with the conditions at the foot of the mountain, but they might also spread to the higher levels, which have warmed. One such example is the scale-backed antbird, an unremarkable little bird, which cannot fly particularly well and hops about hunting army ants, spiders, and snakes.

Once we reach 2,500 feet, Angél and Kevin stop and use their machetes to chop away at ferns and bushes to create a clearing where we can pitch our tents. Stingless bees have commandeered our water canister. Bromeliads spread their leaves like funnels, orchids are blooming, and the air around us smells now sweet, now stale, like a beast of prey. During the night, water begins to drip onto our tents.

At dawn, Wiebe lights the way up the mountain path with his headlamp. Soon I see mosses, lichens, and ferns covering every stone, every available piece of wood. There are even mosses hanging between the vines. Mist rises. A few hundred yards up, Wiebe stops in front of me and listens to a call that is swelling slowly out in the forest: *Rooo-doo-doo-doo-doo!* He turns and looks at me. "A scaled antpitta," he says, "It's much higher up than it should be."

Wiebe makes a funnel shape with his hand and holds it to his mouth, imitating the antpitta, a red-bellied bird with a stumpy tail and stilt-like legs. It returns the same call, somewhere just below us, and then again, from farther away. "There's another one just on the cusp of my range of hearing," he says.

Wiebe recaps the scope of the habitat that Fitzpatrick and Freeman identified: altitudes from 2,300 to 2,600 feet in 1985. Just thirty years later, in 2017, the same habitat lay between 550 and 620 feet higher. And what about now, in 2019? Wiebe pulls his smartphone out of his pants pocket, opens the eBird app, and uses GPS to check our altitude: 3,800 feet. "That's quite a substantial difference," he says.

It's not yet clear how bird species are moving their habitats up the mountain. In the case of the scaled antpitta, Wiebe assumes the following: When the young males fledge, they must seek out their own territory, and there is a high probability that this will be in a zone at a higher altitude because these areas now offer temperatures that suit the species.

These mountain birds may possibly also be able to cope with global warming because they are able to regulate their body temperature independently, even if they have to expend more energy in order to do so. But insects are also migrating up into the mountains, as proven by biologists researching moths on Kinabalu in Borneo.[1] Even sedentary species such as antbirds, which typically hide in the undergrowth, have no choice but to follow their prey up the mountain.

As we approach the mountain ridge, which stands at 4,500 feet, space opens up between the trees, and the white sky emerges. The rain turns the path into a water slide. Again and again, it pulls our legs out from under us, and our hands make contact with the loamy soil, which gives off a moist, earthy scent. By the time we reach the ridge, we're soaked to the skin with rain and sweat. The constant moisture in the air makes it even worse: We're freezing as we sneak a first glimpse of the forested neighboring mountain ridge, Teparo Punta.

Once we've had a breather, we hike along the ridge, until Wiebe stops and listens. His mouth hangs open as he turns to me. "That was another antpitta!"

We climb down into a V-shaped recess in the ground. Sandy sediments attest to this once being a riverbed. Up on the opposite slope, a trail of the remains of branches and leaves shows us the way. But once we've climbed the stony, slippery ramp, the path ends. Wiebe looks around, beating his way up through the wet thicket, but that too is completely overgrown. We daren't

The scaled antpitta may be moving to higher ground with each successive generation because of warmer temperatures, which contribute to a drop in insect populations that the birds rely on for food, and the encroachment of rival bird species into their territory.

think about what might be creeping and crawling in the under-growth around us.

We're just 100 feet from the summit, but it stays out of reach, as if the mountain were reluctant to give up its secrets. It wants to keep its mysteries under wraps.

Wiebe can only speculate as to what is happening up there on the mountain's peak. If conditions are no longer suitable, the birds may scatter along the mountain ridge, perhaps as far as the Andes, which it connects to. But it may also be the case that the birds can no longer find enough to eat, feed their young, or lay their eggs, until, eventually, they go extinct. The birds themselves may be driving this process; you can't stand still on an escalator for long, otherwise the people behind will crash into you. Once rivals force their way up from lower altitudes and compete with long-established species for food and nesting sites, natural selection begins. Whoever has adapted best will prevail.

Descending the mountain is a balancing act, each step ventured on wet rock and roots. Once the rain lets up a little, we stop at 3,300 feet, sit on a tree trunk cushioned with moss, and eat turkey escalopes made for us by César. "That was unexpected," Wiebe concludes, referring to the antpitta's upward migration and its new home on the mountain ridge. I ask him if it will make the antpitta a candidate for early extinction.

"Yes," says Wiebe, nodding. Thinking out loud, he describes species that have disappeared from this area within his own lifetime. These species bore the names of warriors and were somewhat eccentric looking, like the hazel-fronted pygmy tyrant, the crested quetzal, the variable antshrike, the buff-browed foliage-gleaner, and the fulvous-breasted flatbill. Wiebe stares into the trees and mutters to himself, "It would be pretty cool if they were still here."

They may still exist elsewhere in the Andes, but they are highly likely to have been wiped out on this mountain. "We expect this kind of thing to happen when a forest is cleared," says Wiebe. "But this whole area is covered in rain forest, and it's practically untouched."

Wiebe goes back over what he has seen. Birds like Carmiol's tanager, the southern nightingale wren, and the scaled antpitta, which don't belong where he spotted them. He searched for other birds but couldn't find them. These included the russet-crowned warbler, which was still frequently to be seen peeking its little head out of scientists' nets just beneath the mountain ridge in 1985. The population has collapsed by 75 percent in the last three decades; those that remain have flown to new habitats just below the summit.

What's happening on the Cordillera del Pantiacolla holds a mirror up to climate change; these are no longer predictions. "You can see it happening right here," says Wiebe, lifting both his hands and looking up at the tree canopy. "Right now."

ATLANTIC OCEAN

MANÚ
NATIONAL
PARK

PANTANAL

IUCN Protected Area

16

From Rain Forest to Savanna

I t may sound odd but, despite all the adversities they face, the birds, insects, and trees in Manú National Park are comparatively comfortable. The park lies on a slope, running from the Amazonian lowlands 13,000 feet up to the snow-capped peaks of the Andes. Rain forest, mountain rain forest, high mountain ranges. If you need to flee to cooler climes, having a tropical mountain on your doorstep makes for a much quicker trip. "It's a privilege to have the option of migrating up into the mountains," says Gunnar Brehm from the Institute for Zoology and Evolutionary Biology at Friedrich Schiller University, Jena, Germany.

In the lowlands, animals and plants must cover much greater distances to reach cooler regions in the north or the south. In the mountains, temperatures drop by 3°C (5.4°F) with every 1,640 feet of elevation, but in the lowlands, this only occurs with every 310 *miles* of poleward travel.[1] The situation in the tropics is particularly precarious because temperatures around the equator are relatively consistent, as Humboldt demonstrated on his isotherm maps. If you travel three degrees of latitude north of the equator, you will still be surrounded by the same warm air. Only once you leave the tropics do temperatures begin to drop noticeably with each degree of latitude, at a rate of approximately 1°C (1.8°F) every 125 miles. Thus, if a resident of the Amazon strikes out north, it will have to slog its way through the Isthmus of Panama and through the desert as far as Mexico. However, in east Bolivia on the southern end of the

Amazon rain forest, drought and fire are already pushing the rain forest back, and the savanna is growing.[2]

For lowland species, the tropical mountains become their only way out.

But what if there are no mountains for miles around? Take, for instance, the Congo Basin in West Africa, or the region surrounding Manaus in west Brazil, in the heart of the Amazon rain forest, where a flat plain seems to lead on forever. "Species in Manaus have to cover a good 1,250 miles before it turns significantly cooler," says Brehm. "And they have to do that without a specific migration plan, or any conscious decisions."

To put it mildly, escaping climate change in the lowlands is something of a challenge—but it's not completely impossible, as tropical species have repeatedly proven over the course of Earth's history.

MONKEYS IN A TRAP

If the rush for the mountain peaks has already begun, then surely lowland species must also have begun migrating long ago. Over much greater distances, too, in order to reach regions with a similar degree of cooling.

They may well have already begun their migration; it may simply not have been documented yet. There are good reasons why we might fail to spot it: For one thing, scientists long assumed that the climate in tropical rain forests would warm very slowly, and that such studies weren't worth pursuing, especially given more obviously pressing issues, such as deforestation and hunting of wild species. Opinions did not change until 2007, when the science community recognized that climate change would accelerate in the tropics at the same rate as elsewhere.[3] Since research into the area began so late, however, insights and findings are still lagging behind.

On the other hand, researching species migration in the lowlands of the tropics is a costly and unattractive prospect, since the phenomenon is less obvious and also less predictable than it is in the tropical mountains,[4] where it is playing out in a much smaller area. In the lowlands, more than half of all tropical species may simply not be able to travel the necessary and vast distances required in a short space of time to remain in their climate niches, saving themselves by journeying up the nearest tropical mountain. Their range is too small and climate change is advancing too quickly.[5] According to calculations made with the aid of computer modeling by US environmental scientists a few years ago, the same applies to mammals.[6] The scientists surveyed five hundred animal species of this class in North and South America, to see whether they were able to keep pace with climate change. Their analysis revealed that artiodactyls (cloven-hooved animals), armadillos, anteaters, and sloths (!) were best placed to cope. However, over a third of mammals in large areas of the Amazon rain forest could stand condemned, including shrews and our relatives among the apes. By the end of the century, the area of distribution for almost all primates could shrink by an average of two thirds.

When more species die or migrate out of a region due to global warming than move into a region, biologists describe this as "biotic attrition." This is precisely what threatens the low-lying tropics: There is little chance of migration into the area, as it is already the warmest region on Earth; however, species in this region are migrating up into the Andes, and the most heat-sensitive and least mobile species are going extinct.[7] "Since temperatures on Earth have been cooling continuously for 14.5 million years, the lowlands will simply not play host to species that are able to cope with higher temperatures in the future," says Brehm. "This could lead to lowland rain forests losing more and more species as the Earth warms, becoming more susceptible to disturbances from invasive species, and gradually dying off, even in large reserves."

What a horrendous thought. Two questions come to mind: Is there still a way to avoid all this? And could biotic attrition have already begun?

OUT OF CHARACTER

November 2016, Amazon rain forest

Sweat was running down Adriane Esquivel Muelbert's brow; her clothes were sticking. She felt like she was in a steam bath as she and her team of cooks, climbers, and biologists pressed on, miles into the Amazon rain forest. Brazilian-born Esquivel Muelbert is a tropical ecologist at the University of Leeds. Now and then, she saw monkeys climbing the enormous trees, and sometimes she would hear a jaguar, all against the backdrop of the many and varied greens of the forest. "The diversity just blows you away," she enthuses. "There are new species every couple of yards."

But was it possible that there was something not quite right with this tropical rain forest? That something was in motion here, something nobody had noticed yet? This was what Esquivel Muelbert hoped to learn. The team had reached the forest tract with a little help from their GPS devices and had begun to search for marked trees, before leaning ladders against one trunk after another and climbing up to measure the circumference. They did this hundreds of times. And this was just one plot of forest of a total of 106 distributed across half of South America.

Esquivel Muelbert is leading a long-term study unlike any attempted before: Over one hundred biologists have regularly measured the trunks of trees in the lowlands of the Amazon rain forest in the past thirty years. They identified the genus, determined the diameter of the trunk, and collected climate data. The project's critics doubted whether all the effort was really worth it; the biologists would not be able to spot any changes anytime soon, they argued. In fact, they did exactly that. "The forest is already starting to change," says Esquivel

Muelbert. "It's gradually changing in character,"

The weather in the Amazon rain forest has turned extreme. The dry periods are drier, the rainy periods are wetter. In a single decade, the Amazon region has experienced three "droughts of the century"—in 2005, 2010, and 2015–16. In the southern region of the forest ecosystem, the number of periods of rainfall has dropped by a quarter. As a consequence, whole sections of the forest are fading, and the branches that remain attest to the existence of a once-intact rain forest.

Previous droughts were explained away by the fact that the Pacific has warmed, a consequence of El Niño, which returns every five years. But the drought of 2015–16 was different. It was fiercer than ever before and could no longer be explained away by high ocean temperatures off the coast of Peru.[8] Something else must have played a part. Gradually, it became clear that it had something to do with the forest itself.

The complex system of forest, forest-dwelling species, and the water cycle seems to have been thrown out of sync. Moisture-loving trees are dying off due to the new, extreme conditions, while drought-resistant trees try to take their place. Even in areas untouched by humans, Esquivel Muelbert found clearings in the forest, where individual Brazil nut trees and other pioneer trees were quickly growing skyward. Yet these trees, adapted to drought though they are, could not fill the gaps quickly enough.[9] "The ecosystem's response lags behind the speed at which the climate is changing."

The drier it becomes, the more tree species the rain forest loses because they are not adapted to the new conditions. The rain forest is at risk of drying out and losing its function as a protective shield for and life source of many tropical species. And this is not due to climate change alone; according to computer modeling, the Amazon rain forest may cross a historic heat threshold in just a few decades' time. Human beings are doing everything they can to accelerate this process—by clearing and burning the forests.

RIVERS IN THE AIR

The Amazon rain forest is home to thousands of species of trees and is one of the most important ecosystems in the world. It stores vast quantities of carbon dioxide—120 gigatons, comparable to the quantity of greenhouse gases produced by the Earth in five years. It is only able to exist because it supplies itself with rain: Air currents from the Atlantic carry moisture into the Amazon region, and this then falls as rain. If there were no forest, much of this water would simply flow away. However, the trees suck up the water out of the ground using their roots and release some of it back into the air via the pores in their leaves. When this occurs billions of times, as it does in the Amazon rain forest, it forms its own moist atmospheric layer—creating rivers in the air.

Fresh water is constantly rising into the bodies of air that migrate westward as far as the edge of the Andes, where they fall as rain and supply most of the rivers in the Amazon region. This circulatory system supplies moisture to regions as far up as northern Argentina. Three million people living in the vicinity of Bolivia's capital city La Paz benefit from the system; the aerial water supply provides the region with a mild, moist climate, which forms the foundation on which local food systems and livelihoods exist.[10] But it is a fragile system. "Beyond a certain level of deforestation, this cycle will begin to fail," says US biologist Thomas Lovejoy.

On her excursions into the forest, Esquivel Muelbert often spent hours beating her way through the undergrowth to her destination only to find that someone had already been there and chopped the trees down. When this happened, she would turn back right away, taking her blank rainproof notebook with her. Esquivel Muelbert was only interested in areas of forests that have remained untouched by humans. She wanted to examine the influence that climate change was having on the forest, so she was obliged to exclude all other human influences.

It was in the state of Mato Grosso, the center of deforestation in Brazil, where she found it most difficult to find any areas of forest suited to her research. More roads were being cut through the rain forest, fanning out in a herringbone pattern, used for transporting felled timber. The forest is being turned into grazing pasture for cows or soy fields. And that all started before Jair Bolsonaro was elected president of Brazil. On the far-right, Bolsonaro has encouraged cattle and soy farmers to take control of the rain forest, has had environmental workers who resist deforestation intimidated or fired,[11] and is determined to dismantle nature reserves. Rates of deforestation and slash-and-burn clearance have skyrocketed.

Toward the middle of this century, the rain forest could disintegrate into two blocs—one cohesive bloc covering around half of its original range in the northwest, and a severely fragmented bloc in the southeast. Biologists have made predictions as to what will happen to six thousand tree species if their climatic zones shift down by almost 310 miles while clearances destroy possible escape routes and rob the area of its habitat: More than half of the present species would disappear.[12]

FIRE: A NEW PHENOMENON

The more the forest is carved up, the more areas become vulnerable to wind and sun, such that the forest is less capable of protecting itself against drought. Climate change has extended the period of drought that affected the southern Amazon region by about a month, which has made it easier for the forest to catch fire.

Fire is a relatively new phenomenon for broad expanses of the Amazon rain forest. For millions of years, there were almost no fires in the forest at all, until human beings arrived four thousand years ago and torched parts of the forest to create pasture for their cows and to grow crops. "Fire changes the rules of the game for the rain forest," explains US paleobiologist Mark Bush from the Florida Institute of Technology. Even smaller fires can have a devastating effect on the forest and its inhabitants, killing whole populations of ground-nesting birds and permanently changing its composition.

For a long time, flames struggled to make any headway because they would be extinguished in the damp of the rain forest. Climate change is altering this. Heat waves and periods of drought are ensuring that fires can eat their way farther into the forests. During a severe drought in the southeast Amazon, scientists noticed that five times more trees burned than in previous fires. Once branches and leaves have dried out and fallen to the ground, they serve as kindling, while the hotter, drier air makes it easier for flames to jump from tree to tree.[13]

Once fire has raged through a patch of tropical forest, it becomes more vulnerable to another fire. The now-absent canopy of leaves causes the forest to dry out and flammable grasses and herbaceous plants to move in. "It's like adding fuel to the fire," says Bush. "The risk of a second fire is suddenly very high."

A forest needs ten or twenty years to recover from a large fire, but it will never be the same again. It will resemble the cleared forests seen at North American and European latitudes, with low copses that are better armed against fire but contain less wood, store less carbon, and shelter fewer species.[14] The more the rain forest stronghold is reduced to a mosaic of primary and secondary forest, into steppes with stand-alone trees and patches of crop fields and pasture, the more the Earth warms and dries out. Eventually, there is no longer sufficient moisture rising out of the forest to supply the rivers in the air. The water cycle runs into problems, and this encourages droughts, which accelerate the changes in the Amazon. "Slash-and-burn clearance, deforestation, and climate change are generating a negative synergy and could transform vast areas of the rain forest into savanna," Lovejoy warns.

Scientists are no longer concerning themselves with the question of how likely such a situation might be. They are now trying to find out *when* it will emerge. It was long thought that 40 percent deforestation and 3–4°C (5.4–7.2°F) of global warming would prompt this change to take place. But then scientists realized that they hadn't taken all the factors into account.

In early 2018, Tom Lovejoy and the Brazilian climate scientist Carlos Nobre calculated that as little as around 20 percent deforestation would be enough to throw the entire system out of balance, as climate change and slash-and-burn clearance are also fueling the process.[15] And that's not far off from where we are today: The Brazilian rain forest has already lost almost a fifth of its forest area and, in Lovejoy's view, has reached the dangerous threshold at which the water cycle may begin to fail. So, are there signs that it's happening already?

A NEW SYSTEM FLICKERS INTO LIFE

The Pantanal in southwestern Brazil is one of the largest wetlands in the world. It was here where the worst drought for centuries raged in 2020; water levels dropped in rivers and lakes, and an enormous fire burned for weeks, destroying a quarter of the whole ecosystem of forests, islands, and grassland. Many residents of the region have switched from livestock farming to ecotourism and now fear for their livelihoods. The rainy season in the area is now 40 percent shorter, which some scientists, such as Nobre, believe is a result of the "rivers in the air" stalling over the Amazon rain forest.[16]

Elsewhere in Brazil and neighboring countries,[17] grain harvests are failing, drinking water is running low, and energy supplies from hydroelectric power stations are crashing. Pathogens such as the Zika virus are able to spread more easily across vast regions of the continent thanks to hot and dry conditions.[18] These may all be signs of a new system flickering into life.

It is still unclear when exactly the Amazon rain forest will cease to exist in its current form, as we have no analogous climate cases from the past to compare it to, and there are too many unknowns in the equation. "Well before 2100, most of the tropics will be subject to climatic conditions outside the range experienced by any tropical ecosystem on Earth for millions of years," writes biologist Richard Corlett in an essay titled "Climate Change in the Tropics: The End of the World as We Know It?,"[19] published a few years ago. "This 'new tropics' will undoubtedly be different from the one we are familiar with, but we do not yet know enough to predict how different."

In order to counteract the breakdown of the system and thus the disappearance of species, Tom Lovejoy recommends a "margin of safety": reforesting the Amazon rain forest to pull it back from the brink. "I hope the government will soon understand that the Amazon region must be maintained as a whole, integrated system," says Lovejoy.

There is no sign of this, however. Slash-and-burn clearances continued, growing during the coronavirus pandemic, when the authorities were less able to monitor the situation than they had been, and big and small-scale farmers felt emboldened to deforest or burn more and more of the forest[20]—all with the support of the government.[21] In 2021, deforestation in the Amazon rainforest reached its highest pace in fifteen years. At the same time, Brazil has committed to reforesting 75,000 square miles of rain forest by 2030 as part of the Paris Agreement on Climate Change.

Slash-and-burn clearances are enabling rain forest nations such as Brazil to close the gap all the faster, creating climatic conditions that tropical species have not experienced for millions of years. It's like throwing oil on the fire. Many species will be pushed beyond their evolutionary limits without time to react. Even nature reserves won't be able to help them.

THE NIGHTMARE'S COMING TRUE

"Nature reserves are not a panacea," according to Kenneth Feeley and his team of US biologists. They have calculated how the climate in the Amazon rain forest will change by the middle of this century, and have concluded that, in as many as two thirds of all outcomes, the existing climate conditions may be lost forever.[22] The good news for the species that live there is that their climate niche is then very likely to shift toward a different tropical nature reserve. However, in order to reach it, these animals and plants will have to migrate hundreds of miles through unprotected landscapes.

Almost two thirds of rain forests are now situated in a mosaic of pastures and agricultural land, making it practically impossible for species to use these areas as escape routes to cooler habitats.[23] In most cases, forest cover is already insufficient for species to move into areas that may offer a tolerable climate in the future. The horrifying vision that came to the environmental activist Robert Peters in the shower that day thirty-six years ago is now coming true.

And so, these species will have to endure the warming of their habitats. The true meaning of this was seen around the world in 2020, when the equator seemed to catch fire. In Pantanal, Brazil, residents found the bodies of anteaters at the side of the road; they had been trying to save themselves from the inferno. Carbonized crocodiles were found in dried-up riverbeds, along with snakes that had bitten their own tails in their death throes.[24]

In eastern Australia, bushfires raged so hard that a cloud of smoke rose into the stratosphere. Eyewitnesses described the fires as resembling an explosion. The leaves of the eucalyptus tree contain oil and will literally explode when exposed to flames. Firestorms emerged, fanned by the wind; the sound was like a freight train. "The heat was just ridiculous," says Christine Hosking, a koala researcher from the University of

Queensland. When the fires began and the roads were still open, she was making her way from Brisbane to Sydney through an apocalyptic landscape. "It was so loud, so explosive."

She knew that koalas would try to climb up into the tops of the eucalyptus trees—it was usually an effective way of escaping the flames—but the flames shot so high that most of the little marsupials were scorched by the fire. Some that did survive dragged themselves over the fences of Australian gardens and drank desperately from swimming pools or dog bowls. It wasn't uncommon for them to then be eaten by the dogs themselves.

Almost 3 billion animals burned to death, suffocated, or were displaced, including 180 million birds, 143 million mammals, and 51 million frogs.[25] Roughly 100 at-risk species lost the majority of their area of distribution in Australia, even in parts of the tropical rain forest. "Lots of species will go extinct," Australian biologist Lesley Hughes tells me; she was still breathing in the smoke all the way from Sydney. "Their former areas of distribution are becoming uninhabitable."

The individual reasons behind nature's flight from climate change are often hard to make out. Sometimes, a different shift in the year's rhythm means that a species and its host plant are unable to find one another. Sometimes host plants dry out. Sometimes animals and plants are exposed to gradual, steady warming that eventually surpasses their tolerance thresholds. And sometimes it's a case of too much drought, or too much rain. Tracing these problems back to their source is anything but easy, and that's why it is often a topic of discussion in science journals.

But the global fires of 2020 were a different case altogether. The flames saw the source of nature's migration made manifest in the most dramatic of ways, and it was visible to all, thanks to the images beamed by camera teams into living rooms across the world.

The exodus has begun.

PART IV

—

Solutions

Reboot

Headquarters of the UN, New York, October 11, 2021

U N General Secretary António Guterres didn't linger too long over his words of greeting as he stepped up to the lectern and spoke to heads of government and state from across the world. "Excellencies, ladies and gentlemen, we are losing our suicidal war against nature" he began in his Portuguese-accented English. "Our two-century-long experiment with burning fossil fuels, destroying forests, wilderness, and oceans, and degrading the land has caused a biosphere catastrophe."

Like a prosecuting attorney, Guterres listed incidences of our mistreatment of the species with whom we share our planet and pronounced his verdict: "Humanity's reckless interference with nature will leave a permanent record—just as today's scientists study the traces of previous extinctions." Over a million species could become extinct. And the collapse of the ecosystems would hit the poorest the most.

Then he called for a ceasefire.

The video-message was directed to the heads of state of nearly two hundred member-states of the Convention on Biological Diversity (CBD). That day, at the highest political level, they were due to discuss how to stop the planet from hemorrhaging its species diversity—the start of the UN Biodiversity Conference, which took place only virtually because of the ongoing coronavirus crisis. Together with the second part of the UN Biodiversity Conference in 2022 in Kunming, China, it aimed to bring about a new agreement to protect biodiversity.

Its noble vision is that humanity will once again live "in harmony with nature." Guterres names it "the foundations for a permanent peace agreement."

By the year 2030, it is intended that 30 percent of the planet's surface will be protected reserves, more than doubling the current number. Scientists have pushed for this for years without success, but many countries have only taken on their advice in light of the proliferating number of horror stories detailing the terrible state of life on earth. Already in the course of the first part of the Conference in 2021, many nations have committed to the 30 percent target, including the EU, Great Britain, Canada, and California.[1] The hope is that all remaining countries will follow suit in 2022.

Conservationists and biologists are celebrating this as a breakthrough, if long overdue. However, the question that accompanies the new target figure is political dynamite, and it is this that may decide whether our ecosystems, and the species that inhabit them, remain viable in the long term. The question is: What should we save?

"The thirty percent target is really important," says biologist Lee Hannah from Conservation International in the US. "But the real challenge is finding the right places for species that are migrating across the planet."

New nature reserves have previously emerged almost at random, and in places too remote for human beings to access, or which offer only poor ground for crops or grazing. These leftover scraps of land are easy to relinquish to nature. But to truly help species as they migrate, the nature reserves of the future will need to focus less on the demands of human beings and more on the needs of animals and plants. This means that they will be required not just in the locations where species live currently but also in those places to which these species will eventually migrate. As a consequence, these future reserves are in even greater need of protection, whereas some existing reserves may possibly cease to be of importance at a later date if the species that they are intended to protect goes astray.

For some conservationists, it might seem like an unreasonable demand. They have spent their lives trying, at the very least, to protect the refuges that still exist from the pressures exerted by the spread of human beings across the globe. The boundaries surrounding these reserves are almost sacred, and now they're supposed to forget all about them? "Human beings crave stability, order, and predictability," says biologist Pierre Ibisch from the University for Sustainable Development in Eberswalde, Germany, explaining resistance to a dynamic approach to nature conservation in the age of climate change.

We will still desperately need these nature reserves, at least for the time being. In reserves, species are exposed to less stress and have demonstrated greater resistance to climate change than members of the same species outside of reserves. "The first priority is to conserve all those locations where natural habitats still exist," says Hannah.

SAVING WHAT CAN BE SAVED

And so, scientists and conservationists have begun their search for the last refuges for nature in a warming world. In scientific literature, it has been compared to the quest for the Holy Grail. Take the Amazon rain forest, where there are places that might not only be home to a particularly large number of species but where there is also a much lower risk of the land drying out. The northwest of the forest is blessed with good weather and topography. In the future, this region could be damper than anywhere in the Amazon Basin—such that the rain forest will be able to hold its ground. However, it is in this region, of all places, where palm oil plantations are spreading, taking advantage of the rain. They do so entirely legally, as the majority of the forests in this place of refuge are unprotected. "The protection of the Core Amazon should be the foundation of all regional conservation strategies regardless of the dimensions of climate change," argue US biologists Timothy Killeen and Luis Solórzano.[2]

They also consider the transition zones between individual ecosystems, or entire biomes, to be of particular importance, whether these lie between rain forest and savanna, lowlands and mountains, or mountain rain forest and upland grasslands. These climatic border zones often host an entire mosaic of landscapes with vastly different geology, topography, soil conditions, and humidity. Species have sheltered here for thousands of years in all kinds of microrefugia. These cooler havens carry animals and plants through the changing climate like Noah's Ark.

Microrefugia are found all across the world. It makes a difference if a species lives near the coast, a river, or a seepage spring. It makes a difference if it is protected by the forest or if it is exposed to the heat in open meadows and fields. Looking a little closer, we can see that not all forests are made equal, and one meadow is not necessarily like another. On hot days, scientists have observed, butterflies flee from open to more dense vegetation and from lower ground to higher meadows. According to British biologists, these "very local differences in temperatures over distances of centimeters to hundreds of meters are equivalent to the magnitude of an extreme climate change scenario by 2100."[3]

From time to time, the population swaps when warmth-loving species seek out cooler refuges. This is why some biologists prefer to speak of temporary holdouts[4] rather than permanent microrefugia, protecting species through ice ages and temperate interglacial periods. Nevertheless, the experts agree that these little-noticed places will be of key importance in terms of species conservation in the age of climate change.

My father, a botanist, once showed me such a place. It was just 2 kilometers (1.2 miles) from my parents' home, in a state-owned forest in Northern Bavaria, Germany. When I was young, I spent countless hours on my mountain bike, whizzing through the dark, mixed forest, but the refuge it provides for cold-loving species never caught my eye at the time. So, I was particularly surprised when, one sunny autumn day, my father and I followed a sloping gravel path and turned into a dense coniferous forest at the foot of the valley. Our walking boots sank into the moss cushioning the ground. There was also moss matted in the branches of trees, just like in the rain forest. A European common frog hopped off as we approached, a black woodpecker loosed its wailing call. It felt as if I'd lived in the same house for decades and had just discovered a door that led into a room I had never seen before, a room full of secrets.

At night in the forest, the air cools and grows heavy, sinking down. Usually, it would simply continue its journey out of the little valley, but because the course of the stream hardly slopes, resembling a trough, the cold air stays trapped inside it. The spongy moss and the little stream also create cooler and moister conditions than in the surrounding areas, and this has an effect on local species. My father points to a species of moss, wavy-leaved cotton moss, *Plagiothecium undulatum*, which usually grows only in the mountains. A few steps farther along the course of the stream, he notices another cold-loving moss species, greater whipwort, or *Bazzania trilobata*, and later he spots species of ferns, lichens, and fungi that typically grow only in the mountains. They've found a hiding place for themselves here.

Larger-scale islands of cool like these are known as macrorefugia. Several of them can be found in the rain forests of northeastern Australia. One field biologist in this particular region was not content to sit idly by and watch as his favorite tropical species were subjected to climate change, so he came up with a plan for how they might be able to continue to live on Earth in the future. It was a plan that serves as a good template for many other places across the world.

Townsville, 2003

The town on the east coast of Australia is like the great gates to an ancient fort. This is where the tropical rain forest begins, extending 450 kilometers (280 miles) up the coast. It is tiny in comparison to other rain forests on the planet, no bigger than the island of Cyprus; it would hardly be worth mentioning if it weren't for the fact that it trumps the others in age by millions and millions of years.[5]

This rain forest contains remnants of the great forest that covered parts of Australia and the Antarctic between 50 and 100 million years ago, when the two regions were still part of the southern supercontinent Gondwana. Thanks to a favorable constellation of geology, its species were largely protected from the Earth's climatic escapades for a long time. This is why the rain forest on Australia's Great Dividing Range is home to primitive species that exist only here, including the Kuranda tree frog *Ranoidea myola* and the Thornton Peak skink *Calyptotis thorntonensis*.[6]

Stephen "Steve" Williams's office in Townsville on the southern edge of the tropical rain forest is only a stone's throw from the forest proper. It's become something of a second home for Williams, a biologist at James Cook University, and he has long been content studying the strange creatures found inside it. "The problem is that at some point I started wondering what would happen to these species as the climate changed," he explained at a conference in Tasmania in February 2016, where hundreds of specialists in marine and land-based ecosystems, biogeographers and geneticists, paleontologists and modelers had gathered for the first time to research species on the move together.

The first thing Williams did was test what would happen to his "favorite cuddly creature," as he calls it: the Herbert River ringtail possum, a long-tailed possum with a brown-black back and a white breast that spends its days holed up in the hollows of trees. Williams calculated how its habitat would change in

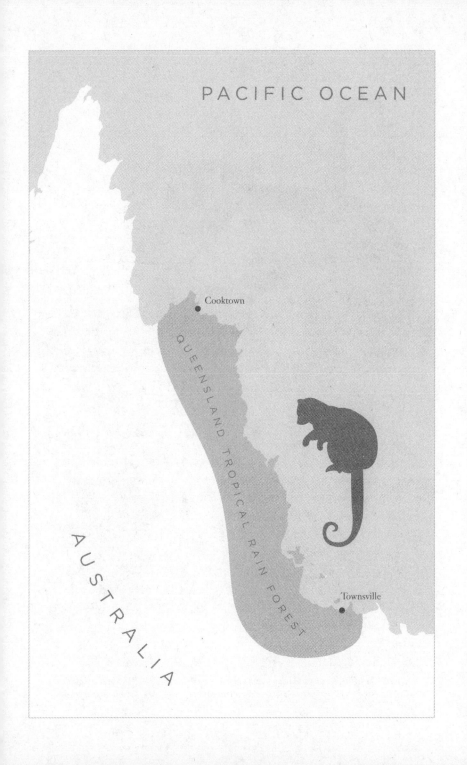

Quickly losing its habitat and facing extinction, the tree-dwelling Herbert River ringtail possum has retreated to higher altitudes in the mountain regions of Queensland, Australia.

various climate scenarios. He determined that, if the world warmed by 3.5°C (6.3°F) over the course of the century—something that seemed quite realistic at the time—inhabitable habitats would shrink until there was almost no habitat in which the possum could survive. Williams carried out the same calculations for other species. "We concluded that half of all endemic species in this world heritage area could potentially go extinct by the end of the century," he explains. Even species that remained would leave the lowlands and retreat into the mountains, where they would retain only 10 percent of their habitat on average. "When I saw the results, the world suddenly seemed gray and depressing."

Most of the species Williams had studied over the course of his career were facing extinction. He couldn't believe it, so he reviewed the modeling and data; he checked whether the animals might be able to migrate away instead, to seek shelter elsewhere, or adapt to another place. Yet the answer remained the same: nature reserves were emptying, at least when it came to the species they were created for, such as the Herbert River ringtail possum. The possum, Williams discovered, had already begun its ascent to higher altitudes.

Other lowland species would be sure to follow, but they tended to be more common and less endangered. The most idiosyncratic creatures on the Australian continent would disappear. Williams outlined this in a study titled "Climate Change in Australian Tropical Rainforests: An Impending Environmental Catastrophe."[7]

In the years that followed, scientists from other regions of the world came to very similar conclusions: Most species' refugia will be uninhabitable by the end of the century. This applies as much to birds in China[8] as it does to elephants in India and Nepal[9] or mammals in Europe.[10] A majority of these will find themselves in a state of climatic disequilibrium. According to ecologists in California,[11] only 8 percent of the world's nature reserves will retain their current climatic conditions beyond the end of this century.

Is there any chance of averting this catastrophe? Williams tested what happened when he factored a milder climate scenario into his modeling.[12] The results showed that only a quarter of all species would be endangered or at risk of extinction. A majority of species would retain part of their habitable habitats—particularly in elevated mountain sites. "It meant we could do something about it," says Williams.

The next step was to find out where the last places of refuge would be at the end of this century, places where the rain forest's endemic species would be able to survive. Williams used computer modeling to determine that these places of refuge were primarily located in the mountains, where the forests provided shade and where cooler temperatures were supplied by the nearby coast. Luckily, 85 percent of these places were already protected. Williams concentrated on those areas that had been degraded and considered where a reforesting project might be of greatest benefit. There were places in the center of the cool refuges, like on the Evelyn Atherton high plateau, which played a central part in protecting species diversity during the ice age cycles of the last 2.6 million years.[13]

Scientists delivered their analysis to the government of Queensland in 2013. And then something unexpected happened: The plan was accepted. The government purchased some of the sites the scientists had determined to be particularly significant, with a view to creating five new national parks, including one that surrounds Mount Baldy near Atherton, and which is now part of the Herberton Range National Park.

THE LAST FIFTY REEFS

Even the ocean has cool places of refuge, where species have a greater chance of surviving climate change. With this in mind, Ove Hoegh-Guldberg, a marine biologist specializing in coral reefs, got to thinking. According to projected figures, a worst-case scenario would see 90 percent of coral reefs disappear by the middle of the century. And yet, he thought to himself, that meant that 10 percent of reefs would make it beyond the year 2050. If we knew where these climate-resistant reefs were, we were obliged to do as much as was humanly possible to protect them. This meant no more overfishing or polluting, and no more overfertilization of the waters that fed them.

Together with his colleagues, Hoegh-Guldberg went on the hunt for the 10 percent. Using computer modeling, he and his team identified fifty sites covering 310 square miles, which they termed "bioclimatic units" (BCUs).[14] These included the waters off Mackay in the heart of the Great Barrier Reef. "Today, the corals in that area are disappearing due to agriculture," Hoegh-Guldberg explains. "But if we stop it, the corals will be able to regrow."

The Great Sea Reef off the islands of Fiji is another example; it is afforded some degree of protection against cyclones and is warming less sharply than other waters, though it remains under threat from land utilization, sugarcane farming, and sea urchins. "Our ambition is to identify these local drivers and then remedy the problem," says Hoegh-Guldberg. This means controlling sea urchins, growing sugarcane sustainably, and reducing the amount of pesticides and fertilizers that reach the seas. "If this experiment works, the newspapers could one day run stories about corals blocking shipping channels because they are thriving so well."

Scientists founded the "50 Reefs" project, joined forces with the WWF, and raised millions of dollars from wealthy sponsors, including former New York mayor Michael Bloomberg. Meanwhile, they have begun putting their plan into practice with six coral reefs, including those off the islands of Fiji, the Solomon Islands, and East Timor. "Everyone wanted to get involved right away, from villagers to prime ministers," says Hoegh-Guldberg. "Many people on the islands of Fiji are devoting their lives to protecting the corals. If we can connect these places to each other, share their enthusiasm, and proceed together, then we'll really have a chance of saving them."

Townsville, 2016

Stephen Williams knew that he still had a way to go. Even the most beautiful reserve was of little use if the species it was intended to protect couldn't reach it. Some future climate refuges, Williams decided, were isolated and were not connected in any way to current reserves. The forests that lay between them had often been cleared, the escape route to sites that would one day be climatically comfortable had been cut off.

What was needed were forest corridors, which species could use to climb to the next higher level. And so, on the advice of Williams and his colleagues, the government of Queensland purchased sites of this kind and began to have them reforested. Meanwhile, even the Australian government took up this approach. It is funding the search for suitable climate refuges and connecting corridors across the country.

But Williams is still not satisfied: He wants to apply his approach to other world regions. For instance, the trapped species of the rapidly fragmenting Amazon rain forest would benefit from emergency exits in the form of new forest corridors. Then they would at least stand a chance of traveling as far from the heat as they could, and, in the meantime, adapting as best they could to the inevitable.[15]

Some tropical countries have already indicated their interest in Williams's approach, including Singapore and Ecuador. They want to build networks for Asia and South America.[16] Many countries in South America and Africa are generally more advanced than industrialized countries (with the exception of Canada[17]), who for a long time did not see the need for a shake-up of their hard-won systems of reserves, simply because they did not consider climate change when creating them. It doesn't matter why a reserve was created, claimed a leading German environmental politician just a few years ago. After all, he claimed, the protection of the most ecologically valuable habitats and open landscape is the best way to support species as they migrate in the era of climate change.[18]

Macroecologists and modelers tested this argument a few years ago. Using species distribution models, they discovered that most species of vertebrates and plants in European nature reserves could be subjected to unsuitable climate conditions by 2080,[19] particularly in the often-compartmentalized Natura 2000 sites, which are sometimes carved up even worse than the unprotected areas that surround them. "Future conservation efforts should be fully aware that the distribution of biodiversity, and species of concern, will be dramatically altered by climate change and that increased extinctions risk is one of the possible outcomes," claim Miguel Araújo and his team from the National Museum of Natural Sciences in Madrid.

The EU's new biodiversity strategy seems to have recognized this. "Protecting the nature we have will not be enough to bring nature back into our lives," explains a communication from the European Commission. Brussels hopes to not simply expand the network of reserves but connect them to each other by means of ecological corridors, "to prevent genetic isolation, allow for species migration, and maintain and enhance healthy ecosystems."[20]

Conflict is inevitable. Ten billion people will need feeding by the middle of the century, but space on Earth is limited. And it is precisely those stretches of land that will play host to a particularly large number of species in the future that will be of interest to human beings in the era of climate change, too. When it gets too hot, farmers will have to move their crops, their fields, and their pastures into cooler regions.[21]

On the other hand, the health and well-being of future generations depends on the continued existence of nature in some intact form. Giving areas over to nature is not an act of altruism. A healthy natural world supplies our cities with water, keeps the ground stable, and prevents floods. It enables us to recuperate and relax. Protecting forests, wetlands, mangroves, and moors is also a form of climate protection, as these ecosystems store vast quantities of CO_2 and methane. Fifteen percent of the greenhouse gases emitted by the planet each year come from natural habitats—most recently, more than half have come from forest clearances in two provinces in Indonesia and Brazil.[22] But it is not just about the forests; it is also a matter of what lives in them. If forest elephants and tapirs disappear from the forests, there will be nothing to eat large, fleshy fruits and distribute their seeds. This would see a reduction in the tallest trees with the greatest wood density, leading to a loss of 10 percent of the carbon stores in the world's tropical forests.[23]

Conversely, the return of the wilderness can have a positive effect on the climate. Following the introduction of conservation measures, the gnu population in Tanzania was able to recover, and the animals kept the vegetation low, leading to fewer fires, and more trees were able to grow, turning the Serengeti from a source of carbon dioxide into a carbon sink. The lesson here is that climate conservation and species conservation are not isolated spheres, where one comes at the cost of the other. They are connected.

THE MATRIX

The EU Commission's vision for making more room for nature includes those areas outside the reserves, known in technical jargon as "the matrix." Agricultural land and cities cannot easily be dismantled to clear the way for species on the move. Often, however, all they need are stepping-stones. Cities should therefore, in Brussels's view, be furnished with woods, parks, and gardens; city farms; green living roofs and walls; promenades, hedges, and meadows, where "excessive mowing is to be avoided." At the same time, a minimum of 10 percent of agricultural land is to be reserved for buffer strips, for fallow land

and hedges, trees, and ponds. Ideally, this would enable species to hop from one refuge to the next.

Conservationists in the western US have developed a flexible solution. They want to turn the Californian longitudinal valley back into a preferred stopover for migrating birds making the journey from South America to the Arctic. Over a period of many years, one wetland after another was replaced with fields, and numbers of these feathery long-distance fliers dropped drastically. Conservationists used birdwatching maps to determine where and when the birds would gather in the remaining wetlands. They rented fields from rice farmers for the duration of the period when the birds would be stopping. The farmers flooded their fields, turning them into wetlands for several weeks. This approach could be applied to other species embarking on one-way journeys.

At the heart of this concept is the notion of reconciling with nature. To stop species extinction, human beings must transform the landscapes they have long dominated, so that they can be used by as many species at once as possible—be this temporary or permanent.[24] Proponents of this approach argue that the return of nature to our landscape and settlements might soothe a deep-rooted yearning for proximity to the natural world. Studies show that people who spend more time in nature are healthier and happier. The science community refers to this as "Vitamin G," where "G" stands for "green space."

We have grown increasingly estranged from nature in the past fifty years. Children, in particular, are having less and less contact with nature in their everyday lives, instead spending more and more time in front of screens.[25] This has consequences: Children exhibit poorer cognitive and motor skills, experience a higher incidence of mental health problems, and place less value on social cohesion. What's worse is that the growing generation no longer recognizes how dependent we are on the natural world, and why we need to protect it. In the science community, this is known as "shifting baseline syndrome." People are steadily lowering their expectations of a healthy environment because they measure the state of nature according to the best experiences they had as children. In other words, they are becoming accustomed to the decline of the natural world.

Conservationists are working to combat this in Great Britain, where they are planning a whole network of pathways of flowering plants to protect pollinators across the country.[26] Those running the initiative hope to cover 370,650 acres with wildflowers. These corridors, each just under two miles wide, are intended to allow wild bees to move back and forth between their isolated habitats as they respond to climate change. "It's important for animals to be able to move from south to north," says Catherine Jones, pollinator officer at the conservation organization Buglife.

Time and again, wild bee conservationists gathered around a table in their office in Peterborough in the east of England to stare at an enormous map of the country. They could see forests, meadows, and heaths, rivers, ponds, and lakes. The activists discussed how best to connect these wild bee habitats to each other and what the most suitable routes for these "insect pathways" might be. They shared suggestions and drew lines. Next, they consulted environmental authorities, the government, city councils, other conservationists, and farmers. "We asked them whether they could transform 10 percent of their land into pollinator-friendly habitats," says Jones.

In the meantime, they have mapped large areas of Great Britain and provided the first 1,200 acres with potential wild bee pathways. Some of these also run through cities—along stepping-stones like parks and gardens. English lawns are to give way to colorful wildflower meadows, where possible fallen branches will not be cleared away, holes in the ground will no longer be filled in, and metal fences will be replaced with hedges. The measures should encourage bumblebees and other pollinators to use these areas to nest and search for food. Anyone who allows their garden to grow into a meadow or plants an apple tree or a currant bush can add this to the map on the Buglife website. "Some people find long grass untidy, or worry that it will attract garbage," says Jones, describing her work in Leeds. "But most people want to get involved."

There are limits, of course. Not all landscapes shaped by human beings can be redesigned to suit other species. And the debate surrounding the reintroduction of wolves, for example, demonstrates that there is a limit to the pleasure many people take in the advance of the natural world. Looking at it from a different angle, many species avoid humans and would never move into a park, for instance, no matter how attractive its redesign might be. These species require vast, unmolested swaths of land where they are free to roam. And nature reserves are still best suited to this—ideally as large and as connected as the draft UN agreement on biodiversity stipulates. It does not necessarily mean keeping people out of as many of these areas as possible. Responsible engagement is an option, and it is possible. Indigenous peoples in the tropics are proof of this. The decline of biodiversity in the areas where they live is less pronounced than elsewhere.[27] Perhaps modern man has forgotten how to engage with nature and needs to be reminded. "We have to see ourselves as part of nature," says Australian biologist Lesley Hughes. "We cannot exist without nature, even if the West likes to imagine that we can."

Even if we protect a substantial portion of the Earth, we would not be able to save all the animals and plants. Even intact landscapes have many species that are unable to migrate because they are simply too slow, and climate change is too fast. "We also have to think about the consequences if species are unable to move their ranges fast enough," says Hughes. "Those that cannot escape or adapt will go extinct."

Unless some divine hand plunges in, picks them up and places them somewhere else.

"ASSISTED MIGRATION"

In 2008, key representatives of the field of climate change biology joined forces to launch an appeal. Writing in the journal *Science*, the group, which included the coral specialist Ove Hoegh-Guldberg, butterfly researcher Camille Parmesan, and Lesley Hughes, called for an end to the taboo: Animals and plants, pushed literally to the brink of extinction by climate change, should, they argued, be resettled in cooler regions where they had not previously existed, as a last resort.

The scientists were quite aware that this strategy of assisted migration would be interpreted as an attack by many conservationists. Many of the values upon which classical conservation is based stem from the intrinsic connection between a species and its original habitat. But now, climate change was posing some uncomfortable questions for the field of conservation: Do we have a responsibility to maintain species diversity? What kind of world do we want to leave to future generations? Should we intervene in nature to save species from extinction, even if doing so means endangering other species? And who gets to decide the answers to questions like these?[28]

As predicted, the call put out by Hoegh-Guldberg, Hughes, and Parmesan was met with considerable resistance. Conservationists accused the group of wanting to "play God."[29] Even their own colleagues warned them of "unintended" and "unpredictable" consequences. Assisted migration of species outside of their natural area of distribution could, ecologists from North America claimed, "create more conservation problems than it solves" and was akin to a game of "ecological roulette."[30] In some extreme cases, introducing new species would only serve to put ecosystems at risk, constraining their viability and spreading diseases. Not to mention the fact that hybridization of local and exotic species would present a major intervention in the evolutionary family tree, no matter how well intentioned.

Hoegh-Guldberg, Parmesan, and Hughes believe the risks to be negligible. Each case must simply be carefully tested and considered. And species must only be resettled within the same continent, and only in regions where the local ecology resembles the species' original habitat. These provisions would make it practically impossible for a species to become invasive. Parmesan, who has moved to a village in the French Pyrenees to continue her butterfly research,[31] recommends looking at the evolutionary history of each species. "If they are only 180 miles apart, it is very likely that they have already interacted with one another in the past several hundred thousand years." Essentially, assisted migration is nothing more than an attempt to connect habitats, just as conservationists have been trying to do for years. "It's not always the right solution," says Parmesan. "But it should at least be an option in the conservationist's toolbox."

Critics argue that even a species that is resettled within a continent and within a specific ecosystem can still wipe out established species and disrupt food chains. It is naïve, they suggest, to believe we can gauge what effects "planned invasions" will have over time or large areas.[32] Not to forget the question of how sensible it might be to pluck out individual elements of an ecosystem and plant them in a new place, while their previous habitat, home to countless species that depend on said elements, has no chance to recover.

"There's still great reluctance to think about moving species," complains Hughes. "Legally, it's a lot easier to clear land and kill animals than it is to move at-risk species to somewhere else."

Nevertheless, it is happening, starting with the Florida torreya, *Torreya taxifolia*. In recent years, this conifer had practically given up its habitat in North America, until there were just a thousand saplings left on the northwestern tip of Florida and in Georgia. In response to this, in 2008 a group of environmentalists calling themselves the Torreya Guardians took it upon themselves to plant thirty-one Florida torreya trees 370 miles farther north in North Carolina[33]—without any kind of administrative supervision, which attracted great criticism.[34]

Then, on August 11, 2016, the first vertebrate to be moved due to climate change was resettled. It was 6 inches in length, with keratotic armor, and is considered to be the rarest reptile in Australia: the Western swamp turtle *Pseudemydura umbrina*. Its habitat had shrunk due to the spread of cities and agricultural land, until there were only about forty of the animals remaining in two wetlands on the edge of the city of Perth. But here, too, conditions were gradually deteriorating as rain failed to fall and the swamps were dry for many months of the year. Even this final place of refuge may become uninhabitable for these creatures by the middle of the century at the latest.

And so, after ten years of preparation, conservation biologists packed two dozen young turtles, bred in captivity, into cardboard boxes, loaded them into an SUV, and drove them 220 miles south of Perth to a cooler and wetter swamp area.[35] There, they carefully released the turtles into the water, and within a few seconds the reptiles had paddled away and gone to explore their new surroundings. The scientists soon lost sight of the well-camouflaged creatures—but transmitters that had been glued to the turtles' shells, along with antennae, provided constant updates on temperature, water depth, and moisture, allowing them to assess whether the new site would be suitable for the endangered turtles in the long term.

Three years later, two species of butterfly were released in Scotland forty miles north of their previous range of distribution and have been spreading across their new habitat ever since.[36] In a mountainous region of southwest China, orchids have been planted 2,000 feet up.[37] There are more plans for a variety of species: Endangered bumblebee species in the Alps or Pyrenees could be resettled in the mountains of Scandinavia, while Arctic mainland species could find places of refuge

The western swamp turtle, the first vertebrate species to be resettled as a result of habitat loss, is the world's most endangered turtle.

on Arctic islands like Spitsbergen or Franz Josef Land.[38] "When we first began writing about it, everyone was saying, 'No! You can't do that!'" Australian biologist Christine Hosking recalls. "But since the situation has worsened so much for wild animals across the world, it's just happening."

Koalas have also been relocated, though this is not yet due to climate change, but because some of their habitats are being forced to give way for new railway lines or houses on the coast of Queensland. The koalas were captured and resettled in the mountains of the subtropical rain forest. "They were doing really well," says Hosking. "But then they were tracked down by a pack of wild dogs and killed."

The Western swamp turtle's new place of refuge was also only a temporary emergency measure. The swampy region in southwestern Australia is still too cold for them, but if climate predictions are correct, in fifty years' time, conditions will be perfect.

This is an issue that affects all species. It is one that drives many conservationists to distraction: All sorts of measures must be taken now to help animals and plants at threat from climate change to move to cooler climes, but only if we know what conditions await them at the end of the century. However, it is hard to predict how quickly human beings will exhaust fossil fuel supplies. It is for this reason that scientists compare conservation in the age of climate change to aiming a gun at a moving target (a slightly odd analogy, admittedly). "The problem with climate change is that there is no end in sight," says Parmesan. "If we knew when the climate would stabilize, we could prepare for that."

But only up to a certain degree of warming. And this brings us to the second problem that climate change presents: If warming crosses a certain threshold, it may no longer be possible to help most species escape it. "It has become clear that beyond 1.5°C (2.7°F), the biology of the planet becomes gravely threatened because ecosystems literally begin to unravel," US scientists warned, writing in *Science* in 2019.[39]

But the problem is that it may soon be impossible to limit warming to 1.5°C. It will only happen if we suck masses of CO_2 out of the atmosphere and push it underground, which could have serious side effects.[40] And so the best-case scenario that remains possible for us is limiting global warming to 2°C (3.6°F) and, in the meantime, giving the world's plant and animal species more space so that they are able to respond to the changing climate and escape to cooler regions, preferably along mountain ranges.

Lee Hannah and his colleagues recently calculated what this would mean for the most species-rich region in the world, the tropics. They found that the risk of extinction would drop by more than half, making it substantially easier to curb.[41]

But it's also true that even if we stop releasing CO_2 into the atmosphere tomorrow, our lethargic planet will continue to warm for at least a decade. Even in the best-case scenario, hundreds of thousands of species would still die out. The majority of these are tropical insects, which are particularly diverse and sensitive to heat. Is it mere coincidence that they are already disappearing from places like Mexico, Puerto Rico,[42] and Costa Rica in droves? These are places where temperatures have already risen considerably or where the dry season has lengthened for several months. Mind you, this has occurred on forested mountains in nature reserves, away from agricultural land. "Where have all the insects gone?" veteran tropical entomologists cry when they roam the rain forest today without being bitten or stung, when they search in vain for spiderwebs but find only flawless leaves, devoid of the slightest bite mark, and feel as if they've walked into a botanical garden.[43] "We've known that we have to do something to combat climate change for forty years, but we haven't really done anything," says Lee Hannah. "We can't just carry on like this and expect there to be no consequences. There *are* consequences."

It won't be an easy thing to accept. There will be a great temptation for apathy in the face of the imminent sixth mass extinction of the Earth—but that would be precisely the wrong response. Because we know, after all, that most species are not helpless when it comes to climate change; they can migrate, if we let them, and because we know the boundaries of this mass exodus, millions of years in the making, we have a clear idea of what is at stake and what we need to do about it. Over the past two decades, hundreds of scientists have shed light on this ancient phenomenon; they have reconstructed past migrations using old fossils and pollen remnants, surveyed the habitats of countless species on every continent, and employed supercomputers to build projections for the future—depending on the choices we make. Their descriptions of the state of the living world do not let us off the hook, but they give us some narrow room for maneuver.

The less we allow the Earth to warm, the more areas we return to nature, and the more reserves and corridors we create, the more species we will be able to save, and we will at least be able to pass on fragments of life on this planet to our children and their children. And if we protect half of the surface of the Earth, which no few biologists have called for,[44] and simultaneously keep global warming below 2°C, Hannah's calculations predict that the risk of extinction would drop by more than three quarters.

"Everyone can be part of the solution," says Australian marine biologist Gretta Pecl. It begins with walking through an environment and keeping an eye out for species that don't actually "belong" there. "People who document the changes they see around them can help us to find out what we can do to protect species."

You want to do more? Great. If you've got a garden, you can make it into an extra inviting pit stop for passing species. If you live near a forest that is due for clearance on less than convincing grounds, you can work to stop it or rewild the area. If you want to make a difference, you can eat less meat and make sure that the meat you do eat is locally sourced. And if you really want to switch things up, you can stop voting for politicians who promise us everything can stay the same, and who don't see climate change for what it is: an existential threat that goes to the very foundations of life on Earth.

And if we give them back the space they need, perhaps animals and plants will surprise us—like the little butterfly at the beginning of this book.

San Ysidro Mountains, at the turn of the millennium

The population of the Edith's checkerspot butterfly *Euphydryas editha* in the far south of California had found itself in an unenviable position. Drought and heat were pushing up from Mexico in the South, leaving the landscape deserted and damaging a small plant with needle-like leaves and transparent flowers, known as *Plantago erecta*: the butterfly's host plant. They could not move north where San Diego, Los Angeles, and the Mojave Desert sprawled, so all they could do was flee upward—as *Euphydryas editha quino*, an endangered subspecies of Edith's checkerspot, had also been proved to be doing. The only problem was that its preferred plant, in the plantain family, would soon no longer grow at this altitude.

Camille Parmesan and her team of biologists were concerned, but also curious to see what would happen. They researched the genetics and behavior of the Quino, as well as its habitat[45]—and made a happy discovery: In the space of fifteen years, the Quino had worked its way up by an average of 360 meters (1,200 feet) to an altitude of 1,164 meters (3,800 feet) and, finding its host plant unavailable, had simply found an alternative. "Within a couple of days, it becomes so desperate that it will accept other plant species," Parmesan explains.

The new plant was *Collinsia concolor*, another plant in the plantain family. The Quino had had little interest in it before and was poorly adapted to it. Nevertheless, the butterfly began to lay its eggs on the plant. When scientists shipped a few samples of the same species from the lowlands into the higher pastures, these butterflies did the same thing. "The lesson is that individuals do not need to be preadapted to novel environments in order to colonize beyond historic range boundaries," says Parmesan. "They merely need to survive."

In doing this, the Quino has bought itself some time. Parmesan estimates that it has a maximum of forty years left on the mountain peak in southern California if we do not curb climate change in the meantime. After this, it will be too hot and too dry for the butterflies. "They can live in new places, but not in new climates."

In this respect, the Quino is not unlike the scaled antpitta in Peru or the Herbert River ring-tailed lemur in Australia. These species with narrow geographical and climatic niches are as at risk of losing their ideal climate as the Arctic cod, the Arctic fox, and the polar bear. Over the course of their evolutionary history, they have always been able to escape to cooler regions when the climate warmed, but now they are at risk of disappearing entirely—and being replaced by species from the lowlands.

In the rain forests of Indonesia, Africa, and the Amazon, on the other hand, earth scientists are expecting to see completely new climate conditions that have yet to take hold on Earth and that howler monkeys, spider monkeys, night monkeys, titi monkeys, capuchin monkeys, and marmosets have not experienced for millions of years.[46] If human beings and their governments do not come to understand them, stop altering the composition of the atmosphere, and liberate them from the hardships they currently endure, then they will all share the same fate: They will be able to endure a certain length of time in a state of disequilibrium with the climate they live in, the most sensitive will tolerate a little less, the more resistant a little more, but eventually, they will all die.

It's practically an iron rule. "We're seeing massive shifts across the whole planet," says Camille Parmesan. "But not a single species is leaving its traditional climatic zone."

Except for one.

—

No More Illusions

Aarhus, Denmark, 2020

Jens-Christian Svenning was staring at his computer screen in his office at the University of Aarhus. It showed nine little blue boxes, each containing a blob; they looked like aerial photos taken by a thermal imaging camera of atolls in the South Seas. Here it was: the human climate niche.

Could it tell us anything about our future?

Svenning, a wiry man with a gray clipped beard, had made a name for himself by modeling changes in the distribution ranges of different species due to climate change and what it means when species are thrown out of balance with the climate to which they have adapted. Svenning proved, for instance, that a whole raft of warmth-loving trees went extinct in Europe at the beginning of the ice age, and that the tree species that endured are consequently poorly adapted to cope with the global warming of today.

But Svenning, the director of the Center for Biodiversity Dynamics in a Changing World, is also interested in human beings. So, he and his colleagues calculated the climate niche of the Neanderthals and discovered that, contrary to popular belief, our ancient ancestors had much preferred to live along the coast of the Mediterranean and not in the cold mountains and plains of Germany and Eastern Europe.

Indeed, this is probably why a Chinese colleague called up Svenning and asked whether he would like to collaborate on a project that aimed to learn whether human beings did have a climate niche and, if so, how it might shift.

It is a little odd: Scientists have examined the climate niches of more than twelve thousand species, but in all that time, they never thought to consider their own. "People are considered as something different," says Svenning. "They are outside the focus of ecologists' minds."

Svenning, a macroecologist by trade, could not understand this, and so he agreed to join forces with his colleague and sound out the typical range of distribution for human beings, just as he had done for insects, birds, and trees before them. He and his colleagues employed statistical methods to transform huge quantities of data on demographics, land use, and climate into historical, current, and future heat maps.

At first, humans seemed to be something of an exception: Over the course of their three-hundred-thousand-year campaign of conquest beginning in Africa and leading out into the furthermost corners of the world, they have adapted to all possible climate conditions. "None of the factors that restrict the distribution of other species like specific habitat requirements and geographical obstacles seem to stop human beings," stated Xu Chi and his team of anthropologists, climate scientists, and ecologists from the University of Nanjing, in a study published in *PNAS* in May 2020.[1]

Soon Svenning had the chart with the atoll blobs in front of him. One axis showed the temperature, and the other displayed atmospheric humidity—both yearly averages. And regardless of whether he looked at the present day, five hundred years ago, or even six thousand years ago, the human population (the blobs) depicted always concentrated around 13°C (55.4°F) and relatively low humidity. "We were all surprised by how consistent this was over such a long timescale," Svenning explains. "Human societies have changed a lot, so we had expected there to be more changes."

And so, the majority of human beings were concentrated in a surprisingly narrow belt that spanned the globe and had done so for at least six thousand years. It was inside this belt that they preferred to live, plant their crops, graze their cattle, and produce all kinds of goods. This core zone of distribution exhibits a narrow temperature window of, on average, 11–15°C (51.8–59.0°F). Translated onto a map of the world, this strip spans the border region between Mexico and the US—it's no surprise that this is where the Quino's checkerspot finds itself surrounded by cities and agricultural land; the strip snakes along across western and southern Europe, the Middle East, and eastern China as far as Japan. "This is the temperate to Mediterranean zone," Svenning explains.

But why did human beings gather in this "realized niche" and not spread themselves more evenly across the globe, across their fundamental niche, which they are climatically capable of tolerating? The authors of the study offer three reasons for this: In the narrow comfort zone, small-scale farmers can work outdoors without suffering from excessive heat or cold. Moderate temperatures are also conducive to elevated mood and good mental health. It is also in this zone where crop fields, grazing animals, and the economy in general exhibit their greatest productivity.

However, after thousands of years of stability, this temperature niche to which we are particularly suited has started to shift due to climate change. In the next fifty years, the authors estimate, those areas where temperatures currently average 13°C (55.4°F) could see temperatures rise to 20°C (68°F), if we continue diligently burning up our fossil fuel reserves. The climate envelope in which we have preferred to establish ourselves is migrating to higher latitudes: to North America, central and eastern Europe, into the Caucasus and northern China.

Yet the real challenges lie in places where it's already hot: along the equator. "Hot," of course, is relative; while the tropics tend to see temperatures averaging between 25 and 27°C (77 and 80.8°F), there are a few regions experiencing extreme drought and temperatures above 29°C (84.2°F). These cover less than 1 percent of the surface of the planet, primarily in the Sahara. Even human beings largely avoid these regions, as surviving them is tough.

The next fifty years could see the ratio change: Forecasts predict that up to 19 percent of the surface of the Earth could be subjected to the kind of heat experienced in the Sahara today. This harsh and life-threatening zone will span the tropics, where the population is growing vigorously, and which is predicted to be home to 3.5 billion people by 2070.

It's not just the climatic conditions that will be hard to bear for the human beings living along the equator; they will also be increasingly isolated. As the number of humans in the tropics continues to grow, the number of other species in the area is dropping rapidly. Schools of fish are migrating to higher latitudes, such that experts predict that fishing in tropical waters will collapse by 40 percent by the middle of the century (and rise by up to 70 percent in the wealthy North[2]); the tropics will lose vast areas of their forests, and with them their water stores; they will lose countless species, on which the cultures of the Indigenous peoples of these regions depend. "Those regions with the strongest climate drivers, with the most sensitive species, and where humans have the least capacity to respond, will be among the most affected," write Gretta Pecl and an international team of researchers in a 2017 study in *Science*.[3]

Climate change is already making for hot, humid conditions in some parts of the world and pushing human beings close to the threshold of their thermal tolerance. The more external temperatures approach human body temperature, the less the body is capable of regulating itself by emitting the excess warmth into its surroundings. We produce sweat through the pores in our skin, which evaporates and cools us. But if humidity is so high that the air is unable to absorb any additional moisture, our bodies will reach their limits. They will still try to pump as much blood as possible away from our cores and into the periphery—our blood vessels dilate, our heartbeat accelerates. For a brief time, this works, but, just like in a sauna or a steam bath, eventually the circulatory system gives up. Anyone who spends a couple of hours outdoors at temperatures of 95°F and at saturated humidity will die.[4]

Ask Svenning how a particular species will respond if its climate niche shifts and he will give you this answer: "If the climate niche is still accessible, we can assume that the species will begin expanding into new regions. This can happen very quickly or very slowly, depending on the species."

The somewhat unpleasant follow-up question is, What on Earth will happen to our species if a third of the future population of the planet lives outside of the climate niche that humans have adapted to over thousands of years? What happens when their crops stop growing (the first signs of this are already present in the tropics[5]), when they wither and burn, when their livestock collapse in the fields, and farmers dare not venture into the outdoors for too long?

Humans have a better chance of adapting than other species because they are able to shape their environments. This is even truer for the humans of the twenty-first century than it was for farmers six thousand years ago. Humans can retreat into air-conditioned homes, breed heat-resistant crops, and develop smart watering systems for greenhouses. "We are special," Lesley Hughes told me when we last spoke on the phone. "But even we have limits."

Svenning and his colleagues anticipate that, eventually, when our planet crosses a critical climate threshold, many human beings will also try to migrate—when they are at risk of starving to death or dying of thirst, and their bodies can no longer take the heat. "If lots of people are exposed to truly extreme, harsh climate conditions such that they struggle to live, people, too, will try to flee," says Svenning.

Humans are a very mobile species. We can cover great distances in a relatively short time. Unlike most other species, we know exactly where we need to go to find cooler places at higher latitudes, where we might be able to lead lives worth striving for in the future.

Let me make it clear: In a worst-case scenario, 3.5 billion people could be living in a barely inhabitable zone by 2070—that's without any reduction in greenhouse gas emissions. "It would be a catastrophe," says Svenning. "We have to avoid it at all costs."

However, in late 2020, something was set in motion—and I'm not talking about the US election. The EU, Great Britain, Canada, South Korea, Japan, and China all affirmed their commitment to more ambitious climate targets and hope to become climate neutral in the coming decades. Climate analysts are no longer ruling out the notion of humanity successfully keeping global warming under 2°C (3.6°F).

Svenning and his colleagues have also calculated how many people might live in the extreme heat zone in 2070 in this scenario. The number is lower (1.5 billion), but it still represents almost a sixth of the anticipated world population. Not all of them will escape or be able to escape. And many who can flee will first stream into their countries' cities, then into neighboring countries, and then into the rich, cooler North. All of this was clear to Svenning, a Dane, as he looked at the blobs on his screen. But it was still a long way off.

Even so, Svenning collated current studies from other parts of the world; naturally, he wanted to be prepared for any questions journalists might put to him. After carrying out some research, he happened upon studies from journals of some renown, which discussed current, if controversial, cases. These were no longer predictions—they were observations.

Svenning read about Mexicans who had migrated to other parts of the country from particularly drought-ridden counties.[6] He read about people in Pakistan who had left their villages for good following unusually severe heat waves and failed harvests.[7] He read about Syria where, in the run-up to the war, the worst drought in 900 years had destroyed crops and livestock and driven 1.5 million citizens into the outskirts of the big cities and contributed to uprising, war, and mass migration.[8] And he read about a long drought in North and West Africa, which was connected to the great numbers of refugees leaving the regions.[9]

There were other examples, too, like Honduras, Nicaragua, and Guatemala, where floods and drought had regularly destroyed the last few years' harvests and cyclones robbed people of their homes. Hundreds of thousands of them fled to the US.[10]

Svenning was aware that some of these studies presented scientific weaknesses or were mere snapshots. There can of course be all kinds of reasons why people migrate: economic, social, and political. People are always moving. Only in the rarest of cases can we say for sure that someone has migrated as a result of climate change.

Rural residents of Guatemala, for instance, may have fled to the cities due to drought, but it's more likely that what drove them from their homes was the misery of gang warfare and criminality. In Syria, the government was partly to blame for the crisis, driving water poverty with their agricultural policy and leaving the rural population in the poorer North to fend for themselves while droughts raged. Good governance can make a difference when it comes down to whether a person will leave their homeland or not, as Syria's neighbor Jordan demonstrates; Jordan was also suffering from a long-lasting, severe drought, but was able to absorb the consequences of it.

Yet the multitude of example cases, and their concurrence, gave Svenning pause. There may well have been other reasons behind people's decisions to flee, and not all instances of extreme weather could be blamed on global warming, but still, that didn't mean that people were not already fleeing climate change, especially since the planet was bearing witness to more migration than it had ever seen before. It was hard to make out individual patterns of movement, yes, but their direction, at least, was clear. People were lining up to join the great march of species across the globe.

Notes

PROLOGUE: IT BEGINS

1. J. Hansen et al. (1981), "Climate Impact of Increasing Atmospheric Carbon Dioxide," *Science*, DOI: 10.1126/science.213.4511.957.

2. K. Achenbach (2003), "Paläobotaniker lieben Pollen," *Scinexx*, scinexx.de. Accessed October 10, 2021.

3. C. Darwin, *The Origin of Species: By Means of Natural Selection, or the Preservation of Favoured Races in the Struggle for Life* (Cambridge University Press, 1876), 339.

4. T. Webb III (1992), "Past changes in vegetation and climate: Lessons for the future," in R. L. Peters, T. E. Lovejoy, *Global Warming and Biological Diversity* (London: Yale University Press, 1992).

5. It's no coincidence that these words resembled those in a thirteen-year-old study in *BioScience*, a copy of which had found its way into Hughes's hands as she prepared for her speech. Robert L. Peters and Joan D. S. Darling compiled the study. It seemed to be fate that she found it, like manna from heaven, Hughes enthused. It was of great use in preparing for her speech and set the course for her burgeoning career.

6. L. Hughes (2000), "Biological consequences of global warming: Is the signal already apparent?," *Trends in Ecology & Evolution*, DOI: 10.1016/S0169-5347(99)01764-4.

7. J. Lenoir et al. (2020), "Species better track climate warming in the oceans than on land," *Nature Ecology & Evolution*, DOI: 10.1038/s41559-020-1198-2.

8. I.-C. Chen et al. (2011), "Rapid Range Shifts of Species Associated with High Levels of Climate Warming," *Science*, DOI: 10.1126/science.1206432.

9. E. S. Poloczanska et al. (2013), "Global Imprint of Climate Change on Marine Life," *Nature Climate Change*, DOI: 10.1038/NCLIMATE1958.

10. And that's putting it mildly. Instead of being restricted, CO_2 emissions continued to increase year on year until 2019. Increasing areas of natural habitat have also disappeared.

CHAPTER 1: HUNTERS

1. AP/Alaska Public Media, "The northernmost US city is now Utqiagvik," *Deutsche Welle*, dw.com. Accessed January 10, 2021.

2. K. Tape et al. (2018), "Tundra be dammed: Beaver colonization of the Arctic," *Global Change Biology*, DOI: 10.1111/gcb.14332.

3. The science community is still debating why. One theory is that beavers, which were almost wiped out in America in the early twentieth century, are recovering thanks to the waning threat of hunting, and recolonizing their old homes—much like snow hares and elk. In the case of the elk, at least, it seems to have been proved that their spread correlates with the increasing height of shrubs, which in turn correlates with warming in the region.

4. K. Tape et al. (2016), "Range Expansion of Moose in Arctic Alaska Linked to Warming and Increased Shrub Habitat," *PLOS*, DOI: 10.1371/journal.pone.0152636.

5. O. Gilg et al. (2012), "Climate change and the ecology and evolution of Arctic vertebrates, *Annals of the New York Academy of Sciences*, DOI: 10.1111/j.1749-6632.2011.06412.x.

CHAPTER 2: HUNTED

1. P. Hersteinsson and D. MacDonald (1992), "Interspecific competition and the geographical distribution of red and Arctic foxes Vulpes vulpes and Alopex lagopus," *Oikos*. DOI: 10.2307/3545168.

2. It is still not clear why Arctic fox numbers have dropped so dramatically despite the species being protected in Nordic countries since the 1940s. Lemmings are being posited as one cause, alongside the advance of the red fox. Lemmings are the Arctic fox's main food source, and the lemming population is also reacting to climate change.

3. A. Landa (2017), "The endangered Arctic fox in Norway—the failure and success of captive breeding and reintroduction," *Polar Research*, DOI: 10.1080/17518369.2017.1325139.

4. A. Rodnikova et al. (2011), "Red fox takeover of arctic fox breeding den: An observation from Yamal Peninsula, Russia," *Polar Biology*, DOI: 10.1007/s00300-011-0987-0.

5. B. von Brackel, "Die Erde hat Kältefrei," *Süddeutsche Zeitung*, sueddeutsche.de. Accessed January 10, 2021.

6. C. Nolan et al. (2018), "Past and future global transformation of terrestrial ecosystems under climate change," *Science*, DOI: 10.1126/science.aan5360.

7. Gilg, "Climate change and the ecology and evolution of Arctic vertebrates."

8. Darwin, *The Origin of Species: By Means of Natural Selection, or the Preservation of Favoured Races in the Struggle for Life* (Cambridge University Press, 1876), 339.

9. "DNA tests confirm hunter shot 'grolar bear,'" CBC News, May 9, 2006.

10. M. Velasquez-Manoff, "Should You Fear the Pizzly Bear?," *The New York Times Magazine*, August 17, 2014.

11. G. Dickie (2017), "On the march: As polar bears retreat, grizzlies take new territory," *The New Humanitarian*, June 28, 2017.

12. E. Fuglei and R. Anker Ims (2008), "Global warming and effects on the arctic fox," *Science Progress*, DOI: 10.3184/003685008X327468.

13. It won't be easy. Any natural catastrophe, like a volcanic eruption, a hurricane, or a flood could completely wipe out an island-dwelling population, and no new Arctic foxes would be able to take their place. Individual populations could also die out as a result of genetic impoverishment. And then there's the matter of food availability. Arctic foxes used to be able to eat their fill of seabird colonies in summer on islands such as Jan Mayen and Bear Island, between the North Cape and Spitsbergen. In winter, however, they lacked an alternative food source, which was why they traveled across the sea ice to other regions such as northeast Greenland or Spitsbergen where there were large Arctic fox populations. But this will cease to be an option if global heating melts the connecting sea ice.

14. Gilg, "Climate change and the ecology and evolution of Arctic vertebrates."

CHAPTER 3: A CHANGE OF REGIME IN THE OCEAN

1. H. Huntington (2020), "Evidence suggests potential transformation of the Pacific Arctic ecosystem is underway," *Nature Climate Change*, DOI: 10.1038/s41558-020-0695-2.

2. S. G. Oliver, "Utqiaġvik whalers still hope to land a bowhead as season wanes," *Anchorage Daily News*, November 6, 2019.

3. "Two killed in Utqiaġvik whaling accident," KTOO, October 9, 2018.

4. S. G. Oliver, "Utqiaġvik finally celebrates first successful bowhead hunt of season," *Anchorage Daily News*, November 20, 2019.

5. In late 2017, neighboring states, together with the EU and China, reached an agreement for a sixteen-year fishing ban. Scientists can use this time to investigate what fish populations actually live in the region, how they are responding to climate change, and what degree of fishing would be reasonable in the region in the future.

6. "2018–2021 Ice seals unusual mortality event in Alaska," NOAA Fisheries, fisheries.noaa.gov. Accessed January 10, 2021.

CHAPTER 4: WHERE ARE THE WHALES?

1. C. Puxley, "Killer whales moving in on polar bear's territory," *Winnipeg Free Press*, January 31, 2012.

2. "Birnirk national historic landmark," National Park Service, nps.gov. Accessed January 10, 2021.

3. Oliver, "Utqiagvik finally celebrates first successful bowhead hunt of season."

4. As told to me by Tero Mustonen, professor at the Department for Geographical and Historical Studies at the University of Eastern Finland.

5. "Economic Valuation and Socio-Cultural Perspectives of the Estimated Harvest of the Beverly and Qamanirjuaq Caribou Herds," InterGroup Consultants Ltd., arcticcaribou.com. Accessed January 10, 2021.

6. C. Mallory, "How will climate change affect Arctic caribou and reindeer?," The Conversation, theconversation.com. Accessed January 10, 2021.

7. J. Monzón et al. (2011), "Climate change and species range dynamics in protected areas," *BioScience*, DOI: 10.1525/bio.2011.61.10.5.

8. J. S. Christiansen (2017), "No future for Euro-Arctic ocean fishes?," *Marine Ecology Progress Series*, DOI: 10.3354/meps12192.

9. J. S. Christiansen et al. (2014), "Arctic marine fishes and their fisheries in light of global change," *Global Change Biology*, DOI: 10.1111/gcb.12395.

CHAPTER 5: THE BREAD-AND-BUTTER SPECIES SWIM AWAY

1. "The Mackerel War," British Sea Fishing, britishseafishing.co.uk. Accessed January 10, 2021.

2. C. Davies, "Fisherman back sanctions against Iceland over mackerel catch," *The Guardian*, January 6, 2013.

3. C. Seidler, "Island bleibt unbeugsam im Makrelenkrieg," *Der Spiegel*, spiegel.de. Accessed February 14, 2020.

4. Iceland's waters have warmed by 1.8–3.6°C (3.2–6.5°F) in the last 20 years alone; see: O. S. Astthorsson et al. (2012), "Climate-related variations in the occurrence and distribution of mackerel (Scomber scombrus) in Icelandic waters," *ICES Journals of Marine Science*, DOI: 10.1093/icesjms/fss084.

5. M. Bernreuther, C. Zimmermann, "Klima und Kabeljau: Fehlt dem Nachwuchs das richtige Futter?," *ForschungsReport*, literatur.thuenen.de. Accessed January 10, 2021.

6. Alfred Wegener Institute for Polar and Marine Research, "Fact Sheet Die Folgen des Klimawandels für das Leben in der Nordsee," awi.de. Accessed January 10, 2021.

7. The acidification of the oceans is another burden for fish to bear; it saps their energy because they have to work harder to compensate for their environment. They then lack the energy they need to grow or reproduce.

8. Over the course of evolution, organisms that depend on each other have grown to develop in harmony with one another over the course of the year. For example, in parts of the Baltic and North Seas, phytoplankton such as diatoms bloom first, at the beginning of the year. When they are producing at full speed, the time comes for zooplankton, such as copepods, to appear. If these then produce masses of larvae, cod larvae—which feed on copepods—are then able to develop. If the ecological calendar of any of these organisms shifts, the whole system may be thrown into disarray. This is what Thünen scientists fear is happening now.

9. Since 1931, commercial fishing vessels have towed what are known as "plankton recorders" behind them. Readings showed that the boundary between groups that prefer warmer, more southerly waters and groups that prefer cooler, more northerly waters has shifted over 680 miles (1,100 km) north. See B. Keim, "Climate change caused radical North Sea shift," *WIRED*, October 28, 2009.

10. Three times per year since 1992, marine scientists have sailed from along the coast of Portugal to Scotland and from Denmark to northern Norway to count mackerel eggs. An analysis of the data collected showed that, over the past thirty

years, mackerel have moved their spawning grounds northward at a rate of 10 miles (16 km) per decade, due to global warming. During this time, however, their thermal niche has migrated northward by a total of 112 miles (180 km). Scientists explain this incongruity with the fact that there may be a lag in the mackerel's response to warming, or they may adapt to the change in their environment by migrating to their spawning grounds earlier when the waters are still cool. See A. Bruge et al. (2016), "Thermal niche tracking and future distribution of Atlantic mackerel spawning in response to ocean warming," *Frontiers in Marine Science*, DOI: 10.3389/fmars.2016.00086.

11. N. K. Dulvy et al. (2008), "Climate change and deepening of the North Sea fish assemblage: a biotic indicator of warming seas," *Journal of Applied Ecology*, DOI: 10.1111/j.1365-2664.2008.01488.x.

12. L. A. Rogers et al. (2019), "Shifting habitats expose fishing communities to risk under climate change," *Nature Climate Change*, DOI: 10.1038/s41558-019-0503-z.

13. T. Young et al. (2019), "Adaptation strategies of coastal fishing communities as species shift poleward," *Marine Science*, DOI: 10.1093/icesjms/fsy140.

14. K. Pierre-Louis, "Warming waters, moving fish: How climate change is reshaping Iceland," *The New York Times*, December 3, 2019.

15. A diplomatic offensive was the only thing to provide any relief: Germany received additional fishing quota shares in Icelandic waters as compensation for the cod. But Iceland snubbed the Brits.

16. J. Spijkers et al. (2017), "Environmental change and social conflict: the northeast Atlantic mackerel dispute," *Regional Environmental Change*, DOI: 10.1007/s10113-017-1150-4.

17. "Blockaded fish row boat leaves Peterhead," BBC News, August 17, 2010.

18. M. L. Pinsky et al. (2018), "Preparing ocean governance for species on the move," *Science*, DOI: 10.1126/science.aat2360.

19. Gerd Kraus, Thünen Institute, PowerPoint presentation.

20. W. W. L. Cheung et al. (2012), "Review of climate change impacts on marine fisheries in the UK and Ireland," *Aquatic Conservation*, DOI: 10.1002/aqc.2248.

21. M. Kurlansky, *Cod: A Biography of the Fish That Changed the World* (London: Vintage Books, 1999).

22. T. L. Win, "Feature: Iceland reaps riches from warming oceans as fish swim north," Reuters, September 20, 2017.

23. Spijkers, "Environmental change and social conflict: the northeast Atlantic mackerel dispute."

24. A. Bruge et al. (2016), "Thermal niche tracking and future distribution of Atlantic mackerel spawning in response to ocean warming," *Frontiers in Marine Science*, DOI: 10.3389/fmars.2016.00086.

25. J. K. Pinnegar et al., "Socio-economic Impacts—Fisheries," *North Sea Region Climate Change Assessment*, eds. M. Quante, F. Colijn (New York: Springer, 2016).

26. C. M. Free et al. (2019), "Impacts of historical warming on marine fisheries production," *Science*, DOI:10.1126/science.aau1758.

27. "Island verzichtet auf EU-Mitgliedschaft," ZEIT Online, zeit .de. Accessed January 10, 2021.

28. "Makrele im Abwärtstrend: Ein weiterer beliebter Speisefisch verliert das MSC-Siegel für nachhaltige Fischerei," Marine Stewardship Council, msc.org. Accessed January 10, 2021.

29. J. Henley, "Iceland accused of putting mackerel stocks at risk by increasing its catch," *The Guardian*, November 21, 2019.

30. J. Barrie, "EU plans to threaten sanctions on Iceland and Greenland as 'mackerel war' looms," *iNEWS*, August 20, 2019.

31. A. Jardine (2019), "Auf der ganzen Welt schmilzt das Eis unaufhaltsam. Was passiert, wenn das Klima kippt?," *Neue Zürcher Zeitung*, nzz.ch. Accessed January 10, 2021.

32. M.-A. Blanchet (2017), "How vulnerable is the European seafood production to climate warming?," *Fisheries Research*, DOI: 10.1016/j.fish-res.2018.09.004.

33. Pinnegar, "Socio-economic Impacts—Fisheries."

CHAPTER 6: IT'S HEATING UP

1. F. Holl (2019), "Alexander von Humboldt und der Klimawandel—Mythen und Fakten," *Internationale Zeitschrift für Humboldt-Studien*, DOI: 10.18443/273.

2. A. Wulf, "Die Vermessung der Natur," *National Geographic* (Germany), July 2019.

3. A. von Humboldt, *Ideen zu einer Geographie der Pflanzen nebst einem Naturgemälde der Tropenländer* (Tübingen, Germany: F. G. Cotta, 1807).

4. Editorial (2019), "Humboldt's Legacy," *Nature Ecology & Evolution*, DOI: 10.1038/s41559-019-0980-5.

5. Humboldt, *Ideen zu einer Geographie der Pflanzen nebst einem Naturgemälde der Tropenländer.*

6. A. Wulf, *The Invention of Nature* (London: Penguin, 2015).

7. Humboldt, *Ideen zu einer Geographie der Pflanzen nebst einem Naturgemälde der Tropenländer.*

8. A. von Humboldt, *Kosmos. Entwurf einer physischen Weltbeschreibung*, vol. 1 (Tübingen, Germany: F. G. Cotta, 1845).

9. F. Holl (2019), "Alexander von Humboldt und der Klimawandel—Mythen und Fakten," *Internationale Zeitschrift für Humboldt-Studien*, DOI: 10.18443/273.

10. Not even 1 percent of all known species in the world have been mapped to date. And though there are vast quantities of data for certain regions in the world, such as bird-loving nations like Finland or Great Britain, there is almost a complete absence of long-term data in Africa. There is a particularly large amount of data for some taxonomic groups (birds, fish, and flowers) and considerably little for others (bacteria, nematodes, fungi, and algae).

11. J. Lenoir et al. (2020), "Species better track climate warming in the oceans than on land," *Nature Ecology & Evolution*, DOI: 10.1038/s41559-020-1198-2.

12. A glance at the history of Earth reveals that many species were also quite capable of adapting to the new conditions in the temperate and glacial periods. Yes, many animals and plants that live on the planet today were first shaped by earlier climatic changes. Today, however, climate change is happening too fast for most species to adapt.

13. Heat is also transmitted twenty-five times better in water than in air.

14. However, climate buffers like forests can only compensate for warming until they themselves disappear because they are no longer able to tolerate drought or because humans cut them down. It is then that the slowness of many land-dwelling species becomes apparent, and their survival strategies, which are more oriented toward adapting than escaping, become an existential issue.

CHAPTER 7: THE FORESTS ARE ON THE MOVE

1. S. N. Matthews, L. R. Iverson (2017). "Managing for delicious ecosystem service under climate change: can United States sugar maple (*Acer saccharum*) syrup production be maintained in a warming climate?," *International Journal of Biodiversity Science, Ecosystem Services & Management*, DOI: 10.1080/21513732.2017.1285815.

2. J. M. Rapp et al. (2019), "Finding the sweet spot: Shifting optimal climate for maple syrup production in North America," *Forest Ecology and Management*, DOI: 10.1016/j.foreco.2019.05.045.

3. Matthews et al., "Managing for delicious ecosystem service under climate change: can United States sugar maple (*Acer saccharum*) syrup production be maintained in a warming climate?"

4. C. W. Woodall et al. (2009), "An indicator of tree migration in forests of the eastern United States," *Forest Ecology and Management*, DOI: 10.1016/j.foreco.2008.12.013.

5. Matthews et al., "Managing for delicious ecosystem service under climate change: can United States sugar maple (*Acer saccharum*) syrup production be maintained in a warming climate?"

6. A. O. Conrad et al. (2020), "Threats to Oaks in the Eastern United States: Perceptions and Expectations of Experts," *Journal of Forestry*, DOI: 10.1093/jofore/fvz056.

7. M. Hanewinkel et al. (2013), "Climate change may cause severe loss in the economic value of European forest land," *Nature Climate Change*, DOI: 10.1038/NCLIMATE1687.

8. Even the European beech (*Fagus sylvatica*) isn't doing well. This icon of Europe's forests has long been considered the tree species of choice in times of climate change. Without humans, beech forests would cover 60 percent of the country, as they did at the beginning of the Middle Ages. But in July 2018, the leaves of many beech trees in large parts of Germany were already turning brown, and in the summer of the following year half of all beech crowns withered and many deciduous trees died off completely.

9. Hanewinkel, "Climate change may cause severe loss in the economic value of European forest land."

10. There are other methods, all of which have advantages and disadvantages. If, for example, you know the physiological limits of a species, you can determine its ecological niche. And this is the fundamental or absolute niche, not merely the realized or potential niche. Another option is using what are known as dynamic models, which take physical, biophysical, and biogeochemical processes into account and can also simulate what happens when one or several species migrate through a landscape which is overdeveloped or fragmented by farmland, and which simultaneously develops a new climate.

11. R. Pearson, T. P. Dawson (2003), "Predicting the impacts of climate change on the distribution of species: are bioclimate envelope models useful?," *Global Ecology & Biogeography*, DOI: 10.1046/j.1466-822X.2003.00042.x.

12. F. Sittaro et al. (2017), "Tree range expansion in eastern North America fails to keep pace with climate warming at northern range limits," *Global Change Biology*, DOI: 10.1111/gcb.13622.

13. K. Zhu et al. (2011), "Failure to migrate: lack of tree range expansion in response to climate change," *Global Change Biology*, DOI: 10.1111/j.1365-2486.2011.02571.x.

14. M. A. Harsch et al. (2009), "Are treelines advancing? A global meta-analysis of treeline response to climate warming," *Ecology Letters*, DOI: 10.1111/j.1461-0248.200901355.x.

15. S. Delzon et al. (2013), "Field evidence of colonization by Holm Oak, at the northern margin of its distribution range, during the anthropocene period," *PLOS One*, DOI: 10.1371/journal .pone.0080443.

16. K. A. Solarik et al. (2019), "Priority effects will impede range shifts of temperate trees species into boreal forests," *Journal of Ecology*, DOI: 10.1111/1365-2745.13311.

17. W. Vieira et al., "Paying colonization credit with forest management could accelerate northward range shift of temperate trees," forthcoming.

18. J. C. Svenning, B. Sandel (2013), "Disequilibrium vegetation dynamics under future climate change," *American Journal of Botany*, DOI: 10.3732/ajb.1200469.

19. A. Prasad et al. (2020), "Combining US and Canadian forest inventories to assess habitat suitability and migration potential of 25 tree species under climate change," *Diversity and Distributions*, DOI: 10.1111/ddi.13078.

20. P. W. Clark et al. (2022), "Ecological memory and regional context influence performance of adaptation plantings in the northeastern US temperate forests," *Journal of Applied Ecology*, DOI: 10.1111/1365-2664.14056.

21. Probably because of a phenomenon called the arctic amplification and the weakening of the jet stream.

CHAPTER 8: THE INSECTS ADVANCE

1. K. D. Levesque (2021), "The effect of climate change on mountain pine beetle (*Dendroctonus ponderosae Hopkins*) in Western Canada," dissertation.

2. K. R. Sambaraju, D. W. Goodsman (2021), "Mountain pine beetle: an example of a climate-driven eruptive insect impacting conifer forest ecosystems," *CAB Reviews*, DOI: 10.1079/PAVSNNR202116018.

3. C. Robinet, A. Roques (2019), "Direct impacts of recent climate warming on insect populations," *Integrative Zoology*, DOI: 10.1111/j.1749-4877.2010.00196.x.

4. T. J. Cudmore et al. (2010), "Climate change and range expansion of an aggressive bark beetle: evidence of higher beetle reproduction in naïve host tree populations," *Journal of Applied Ecology*, DOI: 10.1111/j.1365-2664.2010.01848.x.

5. K. R. Sambaraju, D. W. Goodsman (2021), "Mountain pine beetle: an example of a climate-driven eruptive insect impacting conifer forest ecosystems," *CAB Reviews*, DOI: 10.1079/PAVSNNR202116018.

6. L. Corbett et al. (2016), "The economic impact of the mountain pine beetle infestation in British Columbia: provincial estimates from a CGE analysis," *Journal of Forest Research*, DOI: 10.1093/forestry/cpv042.

7. C. Lesk et al. (2017), "Threats to North American forests from southern pine beetle with warming winters," *Nature Climate Change*, DOI: 10.1038/nclimate3375.

8. B. A. Jones (2021), "Mountain Pine Beetle Impacts on Health through Lost Forest Air Pollutant Sinks," *Forests*, DOI: 10.3390/f12121785.

9. The historian Timothy Winegard assumes, from the numbers. See K.-M. Mayer (2020), "Die neue Plage," *Focus*.

10. Making 50–100 million people sick.

11. M. U. G. Kraemer et al. (2019), "Past and future spread of the arbovirus vectors *Aedes aegypti* and *Aedes albopictus*," *Nature Microbiology*, DOI: 10.1038/s41564-019-0376-y.

12. They also have increased "vector competence," meaning they are better able to transmit tropical viruses than tiger mosquitoes, as more viruses are able to collect in their saliva.

13. L. Lambrechts et al. (2010), "Consequences of the Expanding Global Distribution of *Aedes albopictus* for Dengue Virus Transmission," *PLOS*, DOI: 10.1371/journal.pntd.0000646.

14. This wasn't always the case. Senegal still plays host to an ancient variety that lays its eggs in tree trunks, and only bites apes and other animals.

15. Kraemer, "Past and future spread of the arbovirus vectors *Aedes aegypti* and *Aedes albopictus*."

16. D. Quammen (2013), *Spillover: Der tierische Ursprung weltweiter Seuchen* (Munich: Pantheon, 2013).

17. C. J. Carlson et al. (2020), "Climate change will drive novel cross-species viral transmission," *bioRxiv preprint*, DOI: 10.1101/2020.01.24.918755.

18. Quammen, *Spillover*.

19. R. M. Beyer et al. (2021), "Shifts in global bat diversity suggest a possible role of climate change in the emergence of SARS-CoV-1 and SARS-CoV-2," *Science of the Total Environment*, DOI: 10.1016/j.scitotenv.2021.145413.

20. Kraemer, "Past and future spread of the arbovirus vectors *Aedes aegypti* and *Aedes albopictus*."

21. "Chikungunya in India," World Health Organization, who.int. Accessed January 10, 2021.

22. M. Amrein (2015), "Das unterschätzte Virus," *Neue Zürcher Zeitung*, nzz.ch. Accessed November 11, 2020.

23. E. Rosenthal, "As earth warms up, tropical virus moves to Italy," *The New York Times*, December 23, 2007.

24. M. Enserink (2008), "A Mosquito goes global," *Science Mag*, DOI: 10.1126/science.320.5878.864.

25. ECDC, "Mission Report Chikungunya in Italy," joint ECDC/WHO visit for a European risk assessment, September 17–21, 2007.

26. M. Soumahoro (2011), "The Chikungunya Epidemic on La Réunion Island in 2005–2006: A Cost-of-Illness Study," *PLOS*, DOI: 10.1371/journal.pntd.0001197.

27. ECDC, "Mission Report Chikungunya in Italy."

28. G. Rezza (2018), "Chikungunya is back in Italy: 2007–2017," *Journal of Travel Medicine*, DOI: 10.1093/jtm/tay004.

29. "Dengue and severe dengue," World Health Organization, who.int. Accessed August 26, 2020.

30. G. La Ruche et al. (2010), "First two autochthonous dengue virus infections in metropolitan France," *Euro Surveillance*.

31. "Epidemiological update: second case of locally acquired Zika virus disease in Hyères, France," European Centre for Disease Prevention and Control, ecdc.europa.eu. Accessed September 3, 2020.

32. A. Gloria-Soria et al. (2021), "Vector Competence of *Aedes albopictus* Populations from the Northeastern United States for Chikungunya, Dengue, and Zika Viruses," *The American Journal of Tropical Medicine and Hygiene*, DOI: 10.4269/ajtmh.20-0874.

33. "New York State's First Known Dengue Fever Infection Found On Long Island," CBS New York, November 20, 2013.

34. "Dengue vaccines," World Health Organization, who.int. Accessed January 10, 2021.

35. "Impfstoff gegen Chikungunya Virus," *Pharmazeutische Zeitung*, pharmazeutische-zeitung.de. Accessed January 10, 2021.

36. E. A. Mordecai et al. (2020), "Climate change could shift disease burden from malaria to arboviruses in Africa," *The Lancet*, DOI: 10.1016/S2542-5196(20)30178-9.

37. S. Ryan et al. (2019), "Global expansion and redistribution of Aedes-borne virus transmission risk with climate change," *PLOS*, DOI: 10.1371/journal.pntd.0007213.

38. However, the authors do not want this to be interpreted as a call to cease any kind of climate protection in the name of global justice, because severe global heating would hit tropical countries especially hard (sea-level rises, megadroughts, hurricanes).

39. Population growth and urbanization also play an important part in this.

CHAPTER 9: THE BUMBLEBEE PARADOX

1. C. Gunkel, "Die vergessene Jahrhundertkatastrophe," *Der Spiegel Online*, spiegel.de. Accessed October 13, 2018.

2. H. M. Hines (2008), "Historical Biogeography, Divergence Times, and Diversification Patterns of Bumble Bees (Hymenoptera: Apidae: Bombus)," *Systematic Biology*, DOI: 10.1080/10635150801898912.

3. P. Rasmont, S. Iserbyt (2012), "The Bumblebees Scarcity Syndrome: Are heat waves leading to local extinctions of bumblebees (Hymenoptera: Apidae: Bombus)?," *Annales de la Société entomologique de France*, DOI: 10.1080/00379271.2012.10697776.

4. This is equivalent to the highest recorded temperature in the Pyrenees; meanwhile, temperatures are approaching a similar point in northern Siberia.

5. Rasmont and Martinet assume that this could be due to species-specific protein structure characteristics. The buff-tailed bumblebee appears to have a large variety of these protein profiles and uses these to adapt its bodily systems to higher or lower temperatures.

6. One such example is the rusty patched bumblebee *Bombus affinis* (taking its name from a certain patch on its fluffy coat). This bumblebee was once widespread in southern Georgia, but now appears only rarely in Illinois, Maine, and Wisconsin, as well as in the mountains of Virginia and West Virginia.

7. V. Devictor et al. (2012), "Differences in the climatic debts of birds and butterflies at a continental scale," *Nature Climate Change*, DOI: 10.1038/nclimate1347.

8. J. T. Kerr et al. (2015), "Climate change impacts on bumblebees converge across continents," *Science*, DOI: 10.1126/science.aaa7031.

9. N. St. Fleur, "Climate Change Is Shrinking Where Bumblebees Range, Research Finds," *The New York Times*, July 9, 2015.

10. P. Donkersley (2020), "Bumblebees in crisis: insect's inner lives reveal what the world would lose if they disappear," The Conversation, February 12, 2020.

11. P. Rasmont, "The high climatic risk of European wild bees and bumblebees," presentation before the European Parliament in Brussels, November 14, 2016.

12. P. Soroye et al. (2020), "Climate change contributes to widespread declines among bumble bee across continents," *Science*, DOI: 10.1126/science.aax8591.

13. J. Bridle and A. van Rensburg (2020), "Discovering the limits of ecological resilience," *Science*, DOI: 10.1126/science.aba6432.

14. "Hummeln," Spektrum der Wissenschaft, spektrum.de. Accessed November 6, 2021.

15. A. M. Klein et al. (2007), "Importance of pollinators in changing landscapes for world crops," Proceedings of the Royal Society B: Biological Sciences, DOI: 10.1098/rspb.2006.3721.

16. A. McGivney (2020), "'Like sending bees to war': the deadly truth behind your almond milk obsession," *The Guardian*, January 8, 2020.

17. S. D. Ramsey (2019), "Varroa destructor feeds primarily on honeybee fat body tissue and not hemolymph," *PNAS*, DOI: 10.1073/pnas.1818371116.

18. B. Rundfunk (2018), "Leih und Wanderbienen auf der Walz," BR-Wissen, br.de. Accessed September 29, 2020.

19. D. R. Artz (2011), "Performance of *Apis mellifera, Bombus impatiens,* and *Peponapis pruinosa (Hymenoptera: Apiae)* as Pollinators of Pumpkins," *Journal of Economic Entomology,* DOI: 10.1603/EC10431.

20. L. A. Garibaldi et al. (2013), "Wild Pollinators Enhance Fruit Set of Crops Regardless of Honey Bee Abundance," *Science*, DOI: 10.1126/science.1230200.

21. J. R. Reilly et al. (2020), "Crop production in the USA is frequently limited by a lack of pollinators," Proceedings of the Royal Society, DOI: 10.1098/rspb.2020.0922.

22. I. Koh et al. (2020), "Modeling the status, trends, and impacts of wild bee abundance in the United States," *PNAS*, DOI: 10.1073/pnas.1517685113.

23. In southern Europe, farmers cultivate many traditional varieties of fruits and vegetables that are dependent on pollination, such as tomatoes and strawberries. However, according to climate predictions, by the middle of the century, vast areas of southern Europe could be so dry that agriculture will no longer be possible.

24. E. Civantos et al., "Potential Impacts of Climate Change on Ecosystem Services in Europe: The Case of Pest Control by Vertebrates," *BioScience*, DOI: 10.1525/bio.2012.62.7.8.

25. D. P. Bebber et al. (2013), "Crop pests and pathogens move polewards in a warming world," *Nature Climate Change*, DOI: 10.1038/NCLIMATE1990.

26. C. A. Deutsch (2018), "Increase in crop losses to insect pests in a warming world," *Science*, DOI: 10.1126/science.aat3466.

27. "Neue UN-Projektionen: Weltbevölkerung wächst bis 2050 auf 9,7 Milliarden Menschen," Deutsche Stiftung Weltbevölkerung, dsw.org. Accessed October 8, 2020.

28. E. Civantos et al., "Potential Impacts of Climate Change on Ecosystem Services in Europe: The Case of Pest Control by Vertebrates," *BioScience*, DOI: 10.1525/bio.2012.62.7.8.

29. P. Rasmont et al. (2015), "Climatic Risk and Distribution Atlas of European Bumblebees," *Biorisk*, DOI: 10.3897/biorisk.10.4749.

30. If, for example, the Arctic bumblebee species *Bombus alpinus* were to vanish from Europe, where it is endemic, it would die out altogether.

31. "Durchführungsverordnung (EU) 2016/1141 der Kommission vom," *Amtsblatt der Europäischen Union*, Die Europäische Kommission, eur-lex.europa.eu. Accessed September 28, 2020.

32. "Invasive Arten: Gefahren der biologischen Einwanderung," World Wildlife Fund, wwf.de. Accessed October 8, 2020.

33. European Commission (2017), "Invasive Alien Species of Union Concern," Luxembourg, Publications Office of the European Union.

34. L. Marion (2013), "Is the Sacred ibis a real threat to biodiversity? Long-term study of its diet in non-native areas compared to native areas," *Comptes Rendus Biologies*, DOI: 10.1016/j.crvi.2013.05.001.

35. Species are classified concerning their origin as follows: There are "native species," which colonized the region of their own accord after the last ice age. Those species that have only managed to do so with help from humans are classified as "alien" or "non-native." This category is then subdivided again, into what are known as archaeobiota (old plants and animals introduced from other continents prior to 1492) and neobiota (plants and animals that were brought here after 1492, the year America was "discovered," and which have increased abruptly due to globalization). The current debate focuses on the latter; some people even speak of a "biological invasion"; see bonn.bund.net.

36. This parrot, with its green feathers, was introduced to Germany from India in the 1960s and, much to the chagrin of many conservationists, spread en masse to gardens and parks—at the expense of other birds that nest in the hollows of trees. Meanwhile, however, wild populations of *Psittacula krameria* are also on their way to Europe, Rasmont claims. They have moved away from their natural range in Pakistan and migrated to Greece via Iran and Turkey and could arrive in Belgium and Germany within a few years.

37. Some scientists, however, see potential benefits for ecosystems. Even so, between 2008 and 2020, hunters in Brittany shot five thousand sacred ibis; the new EU guidelines require them to do so.

CHAPTER 10:
CULTURAL ASSETS UNDER THREAT: JAPANESE KELP

1. G. Pecl et al. (2021), "Climate-driven 'species-on-the-move' provide tangible anchors to engage the public on climate change," forthcoming.

2. B. R. Scheffers, G. Pecl (2019), "Persecuting, protecting or ignoring biodiversity under climate change," *Nature Climate Change*, DOI: 10.1038/s41558-019-0526-5.

3. A. Vergés, A. Sen Gupta, "Sydney's waters could be tropical in decades, here's the bad news . . .," The Conversation, September 15, 2014.

4. N. A. Campbell et al., *Biologie* (Munich: Pearson Studium, 2006).

5. D. Fujita (2010), "Current status and problems of isoyake in Japan," *Bulletin of Fisheries Research Agency*.

6. Y. Nakamura et al. (2013), "Tropical Fishes Dominate Temperate Reef Fish Communities Within Western Japan, *PLOS ONE*, DOI: 10.1371/journal.pone.0081107.

7. D. A. Smale, T. Wernberg (2013), "Extreme climatic event drives range contraction of a habitat-forming species," *Proceedings of The Royal Society B*, DOI: 10.1098/rspb.2012.2829.

8. A. Vergés et al. (2016), "Long-term evidence of ocean warming leading to tropicalization of fish communities, increased herbivory, and loss of kelp," *Proceedings of the National Academy of Sciences*, DOI: 10.1073/pnas.1610725113.

9. A. Vergés et al. (2014), "The tropicalization of temperate marine ecosystems: climate-mediated changes in herbivory and community phase shifts," *Proceedings of the Royal Society B*, DOI: 10.1098/rspb.2014.0846.

10. W. F. Precht, R. B. Aronson (2004), "Climate flickers and range shifts of reef corals," *Frontiers in Ecology and the Environment*, DOI: 10.1890/1540-9295(2004)002[0307:CFARSO]2.0.CO;2.

11. D. J. Booth, J. Sear (2018), "Coral expansion in Sydney and associated coral-reef fishes," *Coral Reefs*, DOI: 10.1007/s00338-018-1727-5.

12. W. F. Precht, R. B. Aronson (2004), "Climate flickers and range shifts of reef corals," *Frontiers in Ecology and the Environment*, DOI: 10.1890/1540-9295(2004)002[0307:CFARSO]2.0.CO;2.

13. H. Yamaho et al. (2011), "Rapid poleward range expansion of tropical reef corals in response to rising sea surface temperatures," *Geophysical Research Letters*, DOI: 10.1029/2010GL046474.

14. Fujita, "Current status and problems of isoyake in Japan."

15. M. Sato, H. Kuwahara (2019), "Introduction of countermeasures against the deforestation of seaweed beds 'Isoyake' in Japan," conference paper.

16. "Global warming wreaks havoc on Japanese edible kelp," *The Japan Times*, March 31, 2020.

17. The natural distribution area of the long-spined sea urchin *Diadema antillarum* lies off the coast of New South Wales, Australia. Over ten years ago, it began to migrate toward Tasmania, where it established itself and spread en masse after the sea in the region warmed, enabling its larvae to survive over winter. This resulted in local kelp forests dying off almost completely. All that remained were deserts of nothing but sea urchins.

CHAPTER 11: A DIRTY LITTLE SECRET

1. A. R. Wallace, *Tropical Nature and Other Essays* (London: Macmillan, 1878).

2. C. J. Reddin et al. (2018), "Marine invertebrate migrations trace climate change over 450 million years," *Global Ecology and Biogeography*, DOI: 10.1111/geb.12732.

3. W. Kiessling et al. (2012), "Equatorial decline of reef corals during the last Pleistocene interglacial," *PNAS*, DOI: 10.1073/pnas.1214037110.

CHAPTER 12: THE CORALS MOVE OUT

1. D. Dunne, "The Carbon Brief Interview: Prof Terry Hughes," *Carbon Brief*, November 22, 2018.

2. T. Hughes et al. (2018), "Spatial and temporal patterns of mass bleaching of corals in the Anthropocene," *Science*, DOI: 10.1126/science.aan8048.

3. C. Del Monaco et al. (2016), "Effects of ocean acidification on the potency of macroalgal allelopathy to a common coral," *Scientific Reports*, DOI: 10.1038/srep41053.

4. IPCC (2018), "Summary for Policymakers," Global warming of 1.5°C. An IPCC Special Report on the impacts of global warming of 1.5°C above pre-industrial levels and related global greenhouse gas emission pathways, in the context of strengthening the global response to the threat of climate change, sustainable development, and efforts to eradicate poverty, ipcc.ch.

5. C. Briggs, "Federal Government spending $2.2m on giant ocean fans in bid to protect Great Barrier Reef," iABC News, December 7, 2020.

6. A. Beattie, P. Ehrlich, *Wild Solutions. How Biodiversity Is Money in the Bank* (New Haven, CT: Yale University Press, 2004).

7. M. Pratchett et al (2008), "Effects of climate-induced coral bleaching on coral-reef fishes—ecological and economic consequences," *Oceanography and Marine Biology: An Annual Review*, eds. R. N. Gibson et al.

8. O. DeSmit (2019), "Pacific Islands face hardships as tuna follow warming waters," Conservation International, conservation.org.

9. "Fish migration due to climate change creates tuna shortage in Fiji," World Wildlife Fund, wwf.panda.org.

10. DeSmit, "Pacific Islands face hardships as tuna follow warming waters."

11. J. D. Bell et al. (2013), "Mixed responses of tropical Pacific fisheries and aquaculture to climate change," *Nature Climate Change*, DOI: 10.1038/NCLIMATE1838.

12. N. N. Price et al. (2019), "Global biogeography of coral recruitment: tropical decline and subtropical increase," *Marine Ecology Progress Series*, DOI:10.3354/meps12980.

13. F. Pearce (2019), "As Oceans Warm, Tropical Corals Seek Refuge in Cooler Waters," Yale Environment 360, e360.yale.edu.

CHAPTER 13: AN ABRUPT CHANGE OF REGIME

1. C. H. Trisos et al. (2020), "The projected timing of abrupt ecological disruption from climate change," *Nature*, DOI: 10.1038/s41586-020-2189-9.

2. D. A. Smale et al. (2019), "Marine heatwaves threaten global biodiversity and the provision of ecosystem services," *Nature Climate Change*, DOI: 10.1038/s41558-019-0412-1.

CHAPTER 14: THE MOUNTAIN FOREST BEGINS TO CLIMB

1. K. J. Feeley et al. (2011), "Upslope migration of Andean trees," *Journal of Biogeography*, DOI: 10.1111/j.1365-2699.2010.02444.x.

2. B. Fadrique et al. (2018), "Widespread but heterogeneous responses of Andean forests to climate change," *Nature*, DOI: 10.1038/s41586-018-0715-9.

3. M. W. Tingley et al. (2012), "The push and pull of climate change causes heterogeneous shifts in avian elevational ranges," *Global Change Biology*, DOI: 10.1111/j.1365-2486.2012.02784.x.

CHAPTER 15: THE ESCALATOR TO EXTINCTION

1. I.-C. Chen et al. (2009), "Elevation increases in moth assemblages over 42 years on a tropical mountain," *PNAS*, DOI: 10.1073/pnas.0809320106.

CHAPTER 16: FROM RAIN FOREST TO SAVANNA

1. A. S. Jump et al. (2009), "The altitude-for-latitude disparity in range retractions of woody species," *Trends in Ecology and Evolution*, DOI: 10.1016/j.tree.2009.06.007.

2. F. E. Mayle et al. (2000), "Millennial-Scale Dynamics of Southern Amazonian Rain Forest," *Science*, DOI: 10.1126/science.290.5500.2291.

3. R. T. Corlett (2011), "Climate change in the tropics: The end of the world as we know it?," *Biological Conservation*, DOI: 10.1016/j.biocon.2011.11.027.

4. There is another reason why this remains difficult to prove. High species diversity in the tropics leads to more competition among species than anywhere else in the world, and this creates its own unique distribution dynamics. When climate change alters the pattern of rainfall or drought, there are winners and losers. Species that had previously been held back by other species are suddenly able to expand when climate change stymies their competition, while they themselves are better able to cope with the new conditions. Instead of retreating, these species will spread.

5. S. J. Wright (2009), "The Future of Tropical Species on a Warmer Planet," *Conservation Biology*, DOI: 10.1111/j.1523-1739.2009.01337.x.

6. C. Schloss et al. (2012), "Dispersal will limit ability of mammals to track change in the Western Hemisphere," *PNAS*, DOI: 10.1073/pnas.1116791109.

7. K. J. Feeley and M. R. Silman (2010), "Biotic attrition from tropical forests correcting for truncated temperature niches," *Global Change Biology*, DOI: 10.1111/j.1365-2486.2009.02085.x.

8. A. Erfanian et al. (2017), "Unprecedented drought over tropical South America in 2016: significantly under-predicted by tropical SST," *Scientific Reports*, DOI: 10.1038/s41598-017-05373-2.

9. A. Esquivel Muelbert et al. (2019), "Compositional response of Amazon forests to climate change," *Global Change Biology*, DOI: 10.1111/gcb.14413.

10. M. Bush (2010), "Nonlinear climate change and Andean feedbacks: an imminent turning point," *Global Change Biology*, DOI:10.1111/j.1365-2486.2010.02203.x.

11. M. Simoes, "Brazil Is Persecuting Its Own Environmental Protection Workers, Whistleblowers Say," Vice, November 30, 2020.

12. V. H. F. Gomes et al. (2019), "Amazonian tree species threatened by deforestation and climate change," *Nature Climate Change*, DOI: 10.1038/s41558-019-0500-2.

13. P. M. Brando et al. (2013), "Abrupt increases in Amazonian tree mortality due to drought–fire interactions," *PNAS*, DOI: 10.1073/pnas.1305499111.

14. J. Barlow, Carlos Peres (2008), "Fire-mediated dieback and compositional cascade in an Amazonian forest, *Philosophical Transactions of the Royal Society*, DOI: 10.1098/rstb.2007.0013.

15. T. E. Lovejoy, C. Nobre (2018), "Amazon Tipping Point," *Science Advances*, DOI:10.1126/sciadv.aat2340.

16. C. Gurk, "Ein Paradies steht in Flammen," *Süddeutsche Zeitung*, October 22, 2020.

17. M. Valente, "Drought-hit Argentina faces water worries amid coronavirus pandemic," Reuters, April 29, 2020.

18. S. Paz, J. C. Semenza (2016), "El Niño and climate change—contributing factors in the dispersal of Zika virus in the Americas?, *The Lancet*, DOI: 10.1016/s0140-6736(16)00256-7.

19. R. T. Corlett (2011), "Climate change in the tropics: The end of the world as we know it?," *Biological Conservation*, DOI: 10.1016/j.biocon.2011.11.027.

20. S. Kirchner (2020), "Corona legt Regenwaldschutz lahm," *Klimareporter*, klimareporter.de. Accessed November 26, 2020.

21. Tagesschau (2020), "Massive Abholzung in der Corona-Krise," tagesschau.de. Accessed January 12, 2021.

22. K. J. Feeley, M. R. Silman (2016), "Disappearing climates will limit the efficacy of Amazonian protected areas," *Diversity and Distributions*, DOI: 10.1111/ddi.12475.

23. R. Senior et al. (2019), "Global loss of climate connectivity in tropical forests," *Nature Climate Change*, DOI: 10.1038/s41558-019-0529-2.

24. Gurk, "Ein Paradies steht in Flammen."

25. "Nearly 3 billion animals killed or displaced by Australian wildfires," *Al Jazeera*, July 28, 2020.

CHAPTER 17: REBOOT

1. S. Sengupta, "Europe moves to protect nature, but faces criticism over subsidizing farms" *The New York Times*, October 23, 2020.

2. T. J. Killeen, L. A. Solórzano (2008), "Conservation strategies to mitigate impacts from climate change in Amazonia," *Philosophical Transactions of the Royal Society B*, DOI: 10.1098/rstb.2007.0018.

3. C. D. Thomas, P. K. Gillingham (2015), "The performance of protected areas for biodiversity under climate change," *Biological Journal of the Linnean Society*, DOI: 10.1111/bij.12510.

4. L. Hannah et al. (2014), "Fine-grain modeling of species' response to climate change: holdouts, stepping-stones, and microrefugia," *Trends in Ecology & Evolution*, DOI: 10.1016/j.tree.2014.04.006.

5. "World's oldest tropical rainforest," Tropical North Queensland, tropicalnorthqueensland.org.au. Accessed January 12, 2021.

6. In 1998, UNESCO designated the region a world heritage site; see UNESCO (2020), "Wet tropics Queensland," whc.unesco.org.

7. S. E. Williams et al., "Climate change in Australian tropical rainforests: an impending environmental catastrophe," *Proceedings of the Royal Society of London B*, DOI: 10.1089/rspb.2003.2464.

8. R. Hu et al. (2020), "Shifts in bird ranges and conservation priorities in China under climate change," *PLOS One*, DOI: 10.1371/journal.pone.0240225.

9. R. Kanagaraj et al. (2019), "Predicting range shifts of Asian elephants under global change," *Diversity and Distribution*, DOI: 10.1111/ddi.12898.

10. M. B. Araújo et al. (2011), "Climate change threatens European Conservation areas," *Ecology Letters*, DOI: 10.1111/.j.1461-0248.2011.01610.x.

11. S. R. Loarie et al. (2009), "The velocity of climate change," *Nature*, DOI: 10.1038/nature08649.

12. The RCP4.5 scenario (4.5 does not stand for a rise in temperature but an increase in radiative forcing) corresponds to warming of 2.6 C (4.7°F) by the end of the century.

13. L. P. Shoo et al. (2011), "Targeted protection and restoration to conserve tropical biodiversity in a warming world," *Global Change Biology*, DOI: 10.1111/.1365-2486.2010.02218.x.

14. H. L. Beyer et al. (2018), "Risk-sensitive planning for conserving coral reefs under rapid climate change," *Conservation Letters*, DOI: 10.1111/conl.12587.

15. K. J. Feeley, E. M. Rehm (2012), "Amazon's vulnerability to climate change heightened by deforestation and man-made dispersal barriers," *Global Change Biology*, DOI: 10.1111/gcb.12012.

16. S. Williams et al., "Let's get serious about protecting wildlife in a warming world," The Conversation, May 28, 2015.

17. Provincial governments have been consulting scientists for years in order to identify and protect future routes for migration; see A. Gonzalez et al. (2019), "Connectivity by design: A multiobjective ecological network for biodiversity that is robust to land use and regional climate change," *Biodiversity and Climate Change*, eds. T. Lovejoy, L. Hannah (New Haven, CT: Yale University Press, 2019).

18. J. Flasbarth (2008), "Klimawandel: Künftige Herausforderungen für den Naturschutz," *NABU: Klimawandel und Biodiversität*, Tagungsdokumentation, September 8, 2008.

19. M. B. Araújo et al. (2011), "Climate change threatens European conservation areas," *Ecology Letters*, DOI: 10.1111/j.1461-0248.2011.01610.x.

20. Europäische Kommission (2020), "Mitteilung der Kommission an das Europäische Parlament, den Rat, den Europäischen Wirtschafts und Sozialausschuss und den Ausschuss der Regionen—EU Biodiversitätsstrategie für 2030—Mehr Raum für die Natur in unserem Leben, Brüssel," eur-lex.europa.eu.

21. L. Hannah et al. (2013), "Climate change, wine, and conservation," *PNAS*, DOI: 10.1073/pnas.1210127110.

22. E. Dinerstein et al. (2019), "A Global Deal For Nature: Guiding principles, milestones, and targets," *Science Advances*, DOI: 10.1126/sciadv.aaw2869.

23. J. P. G. M. Cromsigt et al. (2018), "Trophic rewilding as a climate change mitigation strategy?," *Philosophical Transactions B*, DOI: 10.1098/rstb.2017.0440.

24. M. Rosenzweig (2003), "Reconciliation ecology and the future of species diversity," *Oryx*, DOI: 10.1017/20030605303000371.

25. M. Soga, K. J. Gaston (2016), "Extinction of experience: the loss of human-nature interactions," *Frontiers in Ecology*, DOI: 10.1002/fee.1225.

26. "B-Lines," Buglife, buglife.org.uk. Accessed January 12, 2021.

27. M. De Ferrer, "What can we learn from indigenous groups about safeguarding the environment?," *Euronews*, October 23, 2020.

28. M. W. Schwartz et al. (2012), "Managed Relocation: Integrating the Scientific, Regulatory, and Ethical Challenges," *BioScience*, DOI: 10.1525/bio.2012.62.8.6.

29. S. Goldenberg, "A home from home: saving species from climate change," *The Guardian*, February 12, 2010.

30. A. Ricciardi, D. Simberloff (2008), "Assisted colonization is not a viable conservation strategy," *Trends in Ecology and Evolution*, DOI: 10.1016/j.tree.2008.12.006.

31. At the Institute for Theoretical and Experimental Ecology at the National Research Center in Toulouse, France.

32. A. Ricciardi, D. Simberloff (2008), "Assisted colonization is not a viable conservation strategy," *Trends in Ecology and Evolution*, DOI: 10.1016/j.tree.2008.12.006.

33. Torreya Guardians, torreyaguardians.org.

34. M. W. Schwartz et al. (2012), "Managed Relocation: Integrating the Scientific, Regulatory, and Ethical Challenges, in: *BioScience*, DOI: 10.1525/bio.2012.62.8.6.

35. S. Verhagen, "Australian first to save our rarest reptile," *Australian Geographic*, August 18, 2016.

36. S. G. Willis et al. (2008), "Assisted colonization in a changing climate: a test study using two U.K. butterflies," *Conservation Letters*, DOI:10.1111/j.1755-263X.2008.00043.x.

37. H. Liu et al. (2012), "Overcoming extreme weather challenges: Successful but variable assisted colonization of wild orchids in southwestern China," *Biological Conservation*, DOI: 10.1016/j.biocon.2012.02.018.

38. P. Rasmont et al. (2015), "Climatic Risk and Distribution Atlas of European Bumblebees," *Biorisk*, DOI: 10.3897/biorisk.10.4749.

39. E. Dinerstein et al. (2019), "A Global Deal For Nature: Guiding principles, milestones, and targets," *Science Advances*, DOI: 10.1126/sciadv.aaw2869.

40. In order to achieve this, we would have to cover 18 percent of land with corn fields to allow plants to remove carbon dioxide from the atmosphere. After harvesting, these would then be burned (to generate electricity) and the resulting CO_2 compressed under the ground. The downside of this method is that there would then be far less space for growing food, let alone for animals and plants, which would all be redistributed across the Earth. Some experts believe that this would be more damaging to biodiversity than a 2 degree rise in global temperatures.

41. L. Hannah et al. (2020), "30% land conservation and climate action reduces tropical extinction risk by more than 50%," *Ecography*, DOI: 10.1111/ecog.05166.

42. B. C. Lister, A. Garcia (2018), "Climate-driven declines in arthropod abundance restructure a rainforest food web," *PNAS*, DOI: 10.1073/pnas.1722477115.

43. D. H. Janzen, W. Hallwachs (2019), "Perspective: Where might be many tropical insects?," *Biological Conservation*, DOI: 10.1016/j.biocon.2019.02.030.

44. Degraded areas would have to be rewilded, since more than half of Earth's natural habitats have already been turned into settlements, pasture, and arable land by humans.

45. C. Parmesan et al. (2015), "Endangered Quino checkerspot butterfly and climate change: Short-term success but long-term vulnerability?," *Journal of Insect Conservation*, DOI: 10.1007/s10841-014-9743-4.

46. J. W. Williams et al. (2007), "Projected distributions of novel and disappearing climates by 2100 AD," *PNAS*, DOI: 10.1073/pnas.0606292104.

EPILOGUE: NO MORE ILLUSIONS

1. C. Xu et al. (2020), "Future of the human climate niche," *PNAS*, DOI: 10.1073/pnas.1910114117.

2. W. W. L. Cheung et al. (2009), "Large-scale redistribution of maximum fisheries catch potential in the global ocean under climate change," *Global Change Biology*, DOI: 10.1111/j.1365-2486.2009.01995.x.

3. G. T. Pecl et al. (2017), "Biodiversity redistribution under climate change: Impacts on ecosystems and human well-being," *Science*, DOI: 10.1126/science.aai9214.

4. E. Im et al. (2017), "Deadly heat waves projected in the densely populated agricultural regions of South Asia," *Science Advances*, DOI: 10.1126/sciadv.1603322.

5. W. E. Easterling et al. (2007), "Food, fibre and forest products," *Climate Change 2007: Impacts, Adaptation and Vulnerability. Contribution of Working Group II to the Fourth Assessment Report of the Intergovernmental Panel on Climate Change*, eds. M. L. Parry et al. (Cambridge: Cambridge University Press, 2007).

6. I. Chort, M. de la Rupelle (2015), "Determinants of Mexico-US outwards and return migration flows: A state-level panel data analysis," *Demography*, DOI: 10.1007/s13524-016-0503-9.

7. V. Mueller et al. (2014), "Heat stress increases long-term human migration in rural Pakistan," *Nature Climate Change*, DOI:10.1038/nclimate2103.

8. C. P. Kelley et al. (2015), "Climate change in the Fertile Crescent and implications of the recent Syrian drought," *PNAS*, DOI: 10.1073/pnas.1421533112.

9. C.-F. Schleussner et al. (2016), "Armed-conflict risks enhanced by climate-related disasters in ethnically fractionalized countries," *PNAS*, DOI:10.1073/pnas.1601611113.

10. R. A. McLeman, L. M. Hunter, "Migration and Adaptation to Climate Change," IBS Working Paper, 2009.

Image Credits

Page viii: Courtesy of the Natural History Museum of London Library. From *The Great Butterflies of the Earth: A Systematic Treatment of the So Far Known Large Butterflies*, by Dr. Adalbert Seitz, Alfred Kernen Publisher, 1924. Digitally accessed in the Biodiversity Heritage Library.

Page 21: Courtesy of the Smithsonian Libraries. From the *Annual Report of the New Jersey State Museum*, illustrated by James Audubon and John Bachman, MacCrellish & Quigley, 1908. Digitally accessed in the Biodiversity Heritage Library.

Page 29: Top Image, courtesy of NSCU Libraries, from *Universal Dictionary of Natural History*, by M. Charles d'Orbigny, 1849. Digitally accessed in the Biodiversity Heritage Library. Bottom Image, courtesy of Smithsonian Libraries, from *The Wonders of the Animal Kingdom, Exhibiting delineations of the most distinguished wild animals in the various menageries of this country*, by Robert Huish, published by Williams Clowes, 1829.

Page 39: Public Domain. From Goode, G. B., *The Fisheries and Fishery Industries of the United States*. Section I. Natural History of Useful Aquatic Animals. Washington: Government Printing Office, 1884. Plate 59B.

Page 54: Public Domain. Courtesy of the Freshwater and Marine Image Bank. From *Fishes of North Carolina, North Carolina Geological and Economic Survey, vol. II*, by Hugh M. Smith, Raleigh, NC: E. M. Uzzell & Co., 1907.

Acknowledgments

Writing this book plunged me into a new world I had known literally nothing about just four years before; a world that I was able to tap into, like a course of study. And so, my thanks go first of all to all the scientists who explained their research to me, giving me more of their time than they had to. Special thanks are due to Camille Parmesan and Lesley Hughes, Robert Peters, and Gretta Pecl, who helped me to understand the phenomenon of species on the move, and who amazed me time and again.

I would also like to extend my gratitude to the biologists who invited me to their institutions, and those I was able to accompany on land and on water as they carried out their field work: Norbert Becker, who had me release hundreds of (sterilized) tiger mosquitoes in a district of Ludwigshafen; Marc Hanewinkel, who showed me the reputed forest of the future; Oliver Schweiger, who went on the hunt for bumblebees; Pierre Rasmont, who, despite the COVID-19 crisis, spent two hours at his institute introducing me to the world of bumblebees; the researchers at the Thünen Institute of Sea Fisheries in Bremerhaven, who took me out into the North Sea and explained their work there to me, despite the choppy waters and somewhat unfavorable working conditions.

I would not have been able to travel to the Cordillera del Pantiacolla in Peru without the help of Benjamin Freeman and Alex Wiebe or the support of the Pulitzer Center on Crisis

Reporting or the magazine *Reportagen*, which printed a long piece of reportage on the trip. Christian Weber from the *Süddeutsche Zeitung* provided the impetus for the trip and deserves my heartfelt thanks.

Every chapter has been proofread and fact-checked, and I would like to thank Norbert Becker, Gunnar Brehm, Pierre Ibisch, Hauke Flores, Christopher Woodall, Felix Mark, Carl Schleussner, Oliver Schweiger, Jens-Christian Svenning, and Wolfgang Kiessling for their valuable insights.

I would also like to thank my editor, Sophie Boysen, for an exceptionally pleasant and smooth collaborative experience, for intervening at the right moments ("The book doesn't necessarily need more fish") as well as Nicholas Cizek for his enormous efforts with the English edition. Thanks also to Ayça Türkoğlu for the impeccable translation.

I would especially like to thank my parents, who not only supplied me with cake (complete with tin, by mail) and found the right words of encouragement at just the right moments, but also read the entire manuscript and provided exceedingly helpful insights. Extra thanks are due to my father for his patience when I bombarded him with phone calls at all hours and for helping me to see the forest in my hometown with new eyes. You both helped me scale this mountain.

This book would not have been possible without the support of my wife, Kathrin von Brackel, whose devotion I pushed to the limit, especially in the final weeks. She listened to the most outlandish topics with sincere and touching interest ("Today I spent the whole day working on beech distribution/species distribution models/cod, and you won't believe what I found out . . ."), counterchecked each chapter with her keen eye, and helped me find solutions—which always seemed surprisingly obvious to me—when I was once again close to spreading myself too thin. Thank you!

Index

NOTE: Page numbers in *italics* indicate an illustration or photograph. Page numbers followed by *n* indicate an endnote.

About the Author

BENJAMIN VON BRACKEL is a renowned environmental journalist whose reporting on climate change has appeared in *Süddeutsche Zeitung, Die Zeit,* and *Natur.* He is the cofounder of Klimareporter°, an Environmental Media Prize–winning online magazine dedicated to the climate emergency, and co-author of *Angry Weather: Heat Waves, Floods, Storms, and the New Science of Climate Change.* He is based in Berlin, Germany.